Snailing round the South Seas
The *Partula* story

Justin Gerlach

Snailing round the South Seas
The *Partula* story

Justin Gerlach

Phelsuma Press
2014

Phelsuma Press, Cambridge, U.K.

© Justin Gerlach, 2014
http://islandbiodiversity.com/jg.htm
http://islandbiodiversity.com/phelsumapages.htm

ISBN 978-0-9533787-6-0

Contents

Acknowledgements	vii
Prologue	ix
Chapter 1. *Partula faba* and the explorers	1
Chapter 2. The shell collectors	20
Chapter 3. The conchologist adventurers	37
Chapter 4. The biology of obsession	47
Chapter 5. The Three Professors	69
Chapter 6. The meaning of species	84
Chapter 7. The alien invasion	97
Chapter 8. *Partula* conservation	105
Chapter 9. Raiatea rescue	111
Chapter 10. The nature of the beast	125
Chapter 11. Back in the zoo	134
Chapter 12. Rediscoveries	143
Chapter 13. What does the future hold?	148
References	155
Glossary	166
Index to species	168
Index to people	170

Acknowledgements

I am grateful to Jim Murray, Mike Johnson, Paul Pearce-Kelly and Jean-Yves Meyer for providing photographs, and to Trevor Coote, Ron Gerlach and Laura Gerlach for commenting on the text. Thomas Schiotte enabled me to borrow the Spengler and Chemnitz *Partula* specimens in the Copenhagen University Museum of Zoology.

Prologue

The *Partula* snails of the south Pacific islands first came to my attention in 1981. Like many biologists growing up in the 1970s and 1980s, I found great inspiration in Gerald Durrell's books on his animal collecting exploits and his zoo in Jersey. I have a clear memory of first reading about *Partula* in an article in the 'Dodo Despatch', the junior newsletter of the Jersey Wildlife Preservation Trust. This announced the arrival of *Partula* snails in Jersey and described the fate that had befallen them in Polynesia.

Why this had an impact on me then I do not know; perhaps it was because their decline had been so fast and their extinction so imminent, or because they were different. I had always approved of Jersey zoo's different approach to keeping animals, and conservation – the focus on the obscure small mammals and reptiles. Snails may simply have been the logical extension of that.

Eleven years later I was in Polynesia myself, helping to rescue the very last survivors of some of the remaining *Partula* species. At that time my main focus in working with the snails was their conservation and the ecology of the predators that were eating them to extinction. The evolution of the different species was a sideline interest. After a gap of around 20 years I started compiling the more recent information on their evolution and I realised that these snails have been central to the development of biology from the pastime of rich collectors and curious natural historians to a modern science.

What follows is an account of that development, from the voyages of Captain Cook, through the origins and growth of evolutionary thought, to modern genetics, and the impacts of man and our followers. Even someone who has spent most of a scientific career working with snails must admit that they have their limitations in terms of story-telling. So by necessity this is the story of the remarkable people who came across, and in many cases developed a passion for, a remarkable group of animals.

Chapter 1. *Partula faba* and the explorers

Partula faba

Captain Cook's arrival in Polynesia may not seem an obvious place to begin the story of a snail, but that footfall was the start. Although it was not planned as a search for snails, Cook's first voyage to Polynesia was both the first major scientific expedition and the start of the *Partula* story.

In 1768 an expedition was fitted out to observe the 'Transit of Venus'. Astronomers had realised that by taking precise observations of the path of the planet Venus across the face of the Sun, the distance between the Earth and the Sun could be calculated to a high degree of accuracy. The first recorded observations of a transit in 1631 and its pair in 1639 (they occur twice, eight years apart, every 105.5 or 121.5 years) had enabled the size of Venus to be estimated and the distance between the Earth and the Sun to be calculated very roughly. Edmund Halley proposed that making several observations from widely spaced points on Earth would allow trigonometry to calculate a more accurate distance. Accordingly, for the 1761 transit astronomers from Britain, Austria and France were sent around the world. Their results were patchy as cloud covered much of the transit for most observers; one particularly frustrated observer was Neville Maskelyne who, having travelled to St. Helena, failed to see any of the Transit due to thick cloud. 1769 was to be the last opportunity until 1874, so the Transit of Venus was set to be the greatest scientific event of the 18th century. For the British effort the Royal Society determined that three expeditions were to be sent out: to Hudson Bay, the North Cape and the South Seas.

Maskelyne, now Astronomer Royal, suggested the Marquesas islands or Tonga for the South Seas observation. There was a slight difficulty in these ideas as the Marquesas had not been seen since their discovery in 1595, or Tonga since 1643. If they could not be located, a telescope could not be placed upon them, and without firm ground for the telescope no accurate observations could be made. Nonetheless, the Royal Society asked the king for a grant of four thousand pounds in addition to a suitable ship. The king was persuaded of the value of the enterprise and in March 1768 the 'Endeavour' was brought into the navy and the Admiralty selected James Cook to be her Captain. Cook was a surprising choice, not being a commissioned officer nor having the sorts of influential contacts normally essential to such a promotion. Instead he was selected purely on merit, an unusual occurrence in the

18th century British navy. At the same time the return of the 'Dolphin' commanded by Captain Samuel Wallis reported the discovery of an island ideally situated for the observation: Tahiti. The 'Dolphin' charts would enable the island to be located much more reliably than either the Marquesas or Tonga. The Royal Society seized upon this and informed the Admiralty that Tahiti was to be the new destination, and that the expedition was to be joined by the 26 year old naturalist Joseph Banks.

Banks was affluent, well connected and full of great charm; he had used these attributes to good effect, managing to talk his way onto the expedition in an official capacity. As a result the Transit of Venus expedition became the first British expedition to include a naturalist officially on the ship's books. There had been naturalists on ships in the past, but these were the captain's friends and not official members of the crew. In 1766 Philibert Commerçon had been the world's first official naturalist, on the French vessel 'La Boudeuse', commanded by Louis Antoine de Bougainville. Two years later the tradition of British biological exploration started with Banks, a tradition that was to culminate in Charles Darwin's Beagle voyage and the theory of Evolution by Natural Selection.

Banks brought with him an entourage of eight people, including three remarkably proficient artists. In those days exploration was not a light undertaking,

Fig. 1. Captain Wallis reported the discovery of Tahiti which was ideally situated for observing the Transit of Venus, although the initial reception of the 'Dolphin' had not been friendly. From Hakesworth 1773.

in terms of time and risks. The seriousness of these expeditionary undertakings is indicated by the fact that all three artists died before the voyage returned to Britain. The most important member of the Banks group was Carl Solander, a Swedish botanist, and self-styled medical doctor, notable as a protégée of the great Swedish biologist Carl Linnée (more frequently known in his Latinised version: Carolus Linnaeus). Linnaeus had sent Solander to John Ellis to help in establishing the Linneaus's classification system in Britain:

> "No doubt my much-loved pupil Solander has, ere this, found a tranquil asylum in your friendship. I have recommended him to your protection, as I would my own son...."

When Solander and Banks met by chance Banks was so taken with Solander's enthusiasm and knowledge that he added him to his group of assistants. Solander would provide extra scientific weight to the biological side of the expedition, for he was familiar with Linnaeus's system of naming species. This was spreading rapidly through the rest of Europe, and Banks would soon be fully familiar with its rigour. In the meantime he could rely on Solander to correct his identifications and to name the new species they would find.

Solander had studied natural science under Linnaeus at Uppsala University in 1750. There Linnaeus was an inspirational teacher to a whole generation of European biologists: Tärnström, Kalm, Hasselquist, Torén, Osbeck, Löfling, Rolander, Rolandsson, Adler, Forssåkl, Rothman, Falk, Sparrman, Thunberg, Berlin, Afzelius, Koehler, Alstroemer, Von Troil, Fabricius, and Solander. Whilst most of these may not be household names their contribution to biology is astounding. A very large proportion of the 10,000 or so species described by Linnaeus were collected by these 'apostles' and many of them became great taxonomists in their own rights.

Fabricius described studying with Linnaeus:

> "In winter we lived directly facing his house, and he came to us almost every day, in his short red *robe de chambre*, with a green fur cap on his head and a pipe in his hand. ... His conversation on these occasions was extremely sprightly and pleasant. It either consisted of anecdotes... or in clearing up our doubts, or giving us other kinds of instruction. He used to laugh then most heartily, and displayed a serenity and an openness of countenance, which proved how much his soul was susceptible of amity and good fellowship."
>
> He described lectures, exploration in the countryside, evening games with Linnaeus and his wife, Sundays dancing with the Linnee family. "He was fond of conversation on all subjects relative to natural history... That science almost entirely engrossed his speech, and every thought of his mind..." (original text in Danish).

The concept of classification embodied by the Linnaean system was not really new to Britain. Around 100 years earlier the English natural philosopher John Ray had developed what can be regarded as the first modern classification system. Despite humble origins, as the son of a blacksmith and a herbalist, Ray was educated at Cambridge University and ordained as an Anglican priest. In this role he travelled widely in England, and once to mainland Europe, taking the opportunity to collect natural curiosities. His observations and experimental work became widely respected, and he was elected to the newly-formed Royal Society of London in 1667. His first major scientific achievement was to prove experimentally that the wood of living trees transports water and later he became the first person to produce what can be regarded as a scientific definition of a species. In his 1686 *'History of plants'* he wrote:

> "... no surer criterion for determining species has occurred to me than the distinguishing features that perpetuate themselves in propagation from seed. Thus, no matter what variations occur in the individuals or the species, if they spring from the seed of one and the same plant, they are accidental variations and not such as to distinguish a species... Animals likewise that differ specifically preserve their distinct species permanently; one species never springs from the seed of another nor vice versa".

In modern biological terms he was claiming that species could best be recognised by inherited characters, not those acquired through individual development, so a species is a group of organisms descended from a common ancestry, and by inference are capable of interbreeding. This encompasses both the modern 'cladisitc' and 'biological' species definitions (as discussed in Chapter 6).

Ray's 1660 *'Catalogue of Cambridge Plants'* started his interest in organising the names of species in catalogues. He tried to identify a 'natural system' that would reflect the order of Divine Creation, rather than the existing alphabetical or geographical approaches. His system classified plants upon the characters of their flowers, seeds, fruit and roots. This system became widely adopted in Britain and was picked up by some European natural philosophers. Thus by the time that Linnaeus was established as a prominent naturalist, Ray's natural system was well established. Although this was a great improvement some naturalists found that examining all features of an organism produced a frustrating myriad of different classifications. Linnaeus decided that for plants, at least, reproductive features were of such great importance that the other characters could be disregarded. This new Linnaean system, with its emphasis on the key reproductive characters seemed to promise more stability. It should be noted that the Linnaean system of nomenclature was the specification of the characters that matter, rather than the creation of the Latin names. Latin was still the universal language of the European academics, diplomats and nobles. Accordingly, species were described and named in Latin.

Shortly afterwards French came to replace Latin as the language of diplomacy and English the language of commerce and subsequently all else. Today Latin remains only in a shrinking part of the Catholic church and in scientific names.

In 1758 Solander travelled to England to organise the new British Museum along Linnaean lines. At around this time Solander met Banks and they became firm friends. Under the influence of Solander, Banks planned to go to Uppsala and study with Linnaeus but the possibility of travelling to the South Seas on the 'Endeavour' expedition intervened. Although the main aim of expedition was to measure the Transit of Venus, another significant aspect was exploration. The expedition would also counter any French claims for the Terra Australis Incognita for which Bougainville was searching. When the offer came Banks observed:

> "I immediately told Dr. Solander, who received the news with great enthusiasm, without a moment's delay he promised to give me information about everything pertaining to natural history which might be encountered on such a long and unprecedented voyage. But several days later when we were dining at Lady Monson's table and talking about how I had an unmatched opportunity to enrich science and to become famous, Solander all at once excitedly rose from his chair and asked me with intent eyes: would you like a fellow-traveller. I answered: Someone like you would give me untold pleasures and rewards. Then that is it, he said, I'll travel with you; and from that moment everything was settled and decided."

Ellis described the arrangements to Linnaeus:

> "I must now inform you, that Joseph Banks, Esq. a gentleman of £6000 per annum estate, has prevailed on your pupil, Dr. Solander, to accompany him in the ship that carries the English astronomers to the new discovered country in the South sea, Lat. about 20° South, and Long. between 130° and 150° West from London, where they are to collect all the natural curiosities of the place, and, after the astronomers have finished their observations on the transit of Venus, they are to proceed under the direction of Mr. Banks, by order of the Lords of the Admiralty, on further discoveries of the great Southern continent, and from thence proceed to England by the Cape of Good Hope.... No people ever went to sea better fitted out for the purpose of Natural History, nor more elegantly. They have got a fine library of Natural History; they have all sorts of machines for catching and preserving insects; all kinds of nets, trawls, drags and hooks for coral fishing; they have even a curious contrivance of a telescope, by which, put into the water, you can see the bottom to a great depth, where it is clear. They have many cases of bottles

with ground stoppers, of several sizes, to preserve animals in spirits. They have the several sorts of salts to surround the seeds; and wax, both beeswax and that of the *Myrica*; besides there are many people whose sole business it is to attend them for this very purpose. They have two painters and draughtsmen, several volunteers who have a tolerable notion of Natural History; in short Solander assured me this expedition would cost Mr. Banks ten thousand pounds. All this is owing to you and your writings."

They set sail on 25th August 1768, sailing west, round Cape Horn and into the Pacific, making landfall on Tahiti on 13th April 1769. Banks and Solander were given the charge of restoring their food supplies and this occupation, along with Banks apparently being more taken by the customs of the people than the nature of Tahiti mean that he recorded few observations of his collections. One of his very few notes was: "This morning showery and cool, seemingly a good opportunity of going

Fig. 2. The eminent men associated with Cook's first expedition. Left to right: Daniel Solander, Joseph Banks, Captain Cook, John Hawkesworth (editor of Cook's expedition papers) and John Montague (4th Earl Sandwich, First Lord of the Admiralty). Painting by John Hamilton Mortimer about 1771. Original in the National Library of Australia.

upon the hills. I went accompanied only by Indians, indeed all of them but one soon left me, he however accompanied me during my whole walk. The paths were very open and clear till I came to the woods but afterwards very bad, so much so that I could not reach the top of the lowest of the two high hills seen from the fort, which was all I intended. I was in some measure however recompens'd by finding several plants which I had not before seen, with which I returned before sunset…"

There are no records of any wild animals in Banks's journal. Cook was slightly more forthcoming in passing: on Tahiti "few, either Men or Women, are without a Necklace or String of Beads made of Small Shells or bones about their Necks." These would doubtless have been *Partula* shells, as were used in Polynesian necklaces until the 1980s.

Tahiti was seen as a paradise by the early explorers; Beechey's impressions from the 1826 landing of the 'Blossom' was expanded colourfully by Huish as:

> "The landscape scenery of Otaheite is in general beautiful in the extreme, forming a happy combination of land and water, of precipices and level plains, of trees often hanging their branches, clothed with thick dark foliage, over the sea; and distant mountains

Fig. 3. Early view of Polynesia: Opoa valley, Raiatea, by Daniel Tyerman, 1822. Original in the National Library of New Zealand

shewn in sublime outline and richest hues and the whole often blended in the harmony of nature, produces sensations of admiration and delight. The inland scenery is of a different character, but not less impressive. The landscapes are occasionally extensive, but more frequently circumscribed. There is, however, a startling boldness in the towering piles of basalt, often heaped in romantic confusion near the source or margin of some cool and crystal steam, that flow in silence at their base, or dashes over the rocky fragments that arrest its progress, and there is the wildness of romance about the deep and lovely glens, around which the mountains rise like the steep sides of a natural amphitheatre, till the clouds seem supported by them; this arrests the attention of the beholder, and for a time suspends his faculties in mute astonishment. There is also so much that is new in the character and growth of trees and flowers, irregular, spontaneous, and luxuriant in the vegetation, which is sustained by a prolific soil, and matured by the genial heat of a tropic clime, that is adapted to produce an indescribable effect. When journeying through some of the inland parts of the island, the effect of the scenery, and the unbroken stillness which pervaded the whole were so impressive, that giving full scope to the imagination, the delusion might easily have been induced, that we were walking on enchanted ground, or passing over fairy lands. It has at such seasons appeared, as if we had been carried back to the primitive ages of the world, and beheld the face of the earth as it was, perhaps, often exhibited, when the Creator's works were spread over it in all their endless variety, and all the vigour of exhaustless energy; and before population had extended, or the genus and enterprise of man had altered the aspect of its surface."

By the time Cook's expedition moved to Raiatea island the ship was well provisioned and Banks was more accustomed to the people. He seems to have spent more time in exploration. On 20th July 1769 he recorded in his journal that they landed "at *Ulhietea* [then the name of Raiatea island] in a bay Calld by the natives *Oapoa*, the entrance of which is very near a small Islet Calld *Owhattera*. Some Indians soon came on board expressing signs of fear, they were two Canoes each of which brought a woman, I suppose as a mark of confidence, and a pig as a present. To each of these ladies was given a spike nail and some beads with which they seemd much pleasd." The Polynesians were prepared to trade almost anything for an iron nail, which was considerably harder and more versatile than any of their own materials. This demand for iron was one of the factors that forced expeditions to cut short their visits before all the nails were prised from the ships.

They were keen to land and explore Raiatea as quickly as possible, as according to Banks their guide, a priest called Tupia, "who has always expressd much fear of the men of *Bola Bola* says that they have conquerd this Island and will tomorrow come down and fight with us, we therefore lose no time in going ashore as we are to have today to ourselves." Tupia had joined them on Tahiti where he was effectively a refugee from Raiatea, dispossessed by invaders from Bora Bora island, accordingly he was at pains to make clear to the expedition how dangerous the occupying Bora Bora people were.

After raising the Union flag and taking possession of the island, and the others in sight (Tahaa, Huahine and Bora Bora), Cook, Banks and Dr. Solander toured the settlement. They were most taken by the *Marai* of *Taputapu-atea*, the great temple of Raiatea. The expedition did not realise the significance of what they were looking at for Raiatea was the spiritual homeland of the Polynesians, and this *Marai* the centre of their religion. It was a place of human sacrifice and Banks became intrigued by the bundles of human jaws found in long houses all around the island. "Tupia told us that it was the custom of these Islanders to cut off the Jaw bones of those who they had killd in war; these were he said the jaw bones of Ulhietea people but how they came here or why tied thus to a canoe we could not understand, we were therefore contented to conjecture that they were plac'd there as a trophy won

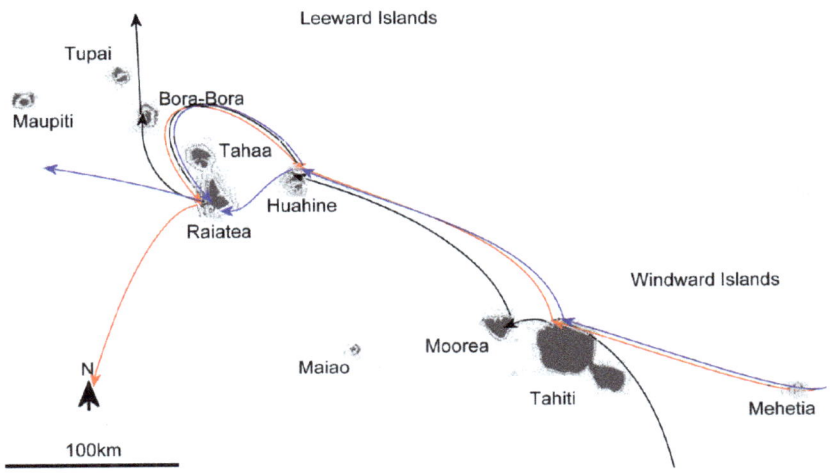

Fig. 4. Map of the Society Islands showing Cook's expeditions: red – 1st expedition; blue – 2nd expedition; black – 3rd expedition

back from the men of Bola Bola [=Bora Bora] their mortal enemies."

Banks explored much of the island over the next two weeks, often in the company of Dr. Solander. His journal however, gives almost no useful detail, although the people interested him the nature of the island was a disappointment, recording "meet little worth observation" or "Dr Solander and myself go upon the hills in hopes of finding new plants but ill rewarded; return home at night having seen nothing worth mentioning." His view is summed up by the 2nd August when Banks and Solander "spent this day ashore and been very agreeably entertaind by the reception we have met with from the people, tho we were not fortunate enough to meet with one new plant." They did explore thoroughly despite their lack of success: "Dr Solander and myself go upon the hills accompanied by several Indians, who carried us by excellent paths so high that we plainly saw the other side of the Island and the passage through which the ship went out of the reef between the Islets of *Opoorooroo* and *Tamou*. Our walk did not turn out very profitable as we found only two plants that we had not seen before."

Finally, on the 4th August they heard that the King of the feared Bora Bora men was intending to pay them a visit. "We are much inclind to receive him civily as we have met with so civil a reception from his subjects." However, the visit did not take place and on the 5th: "Opoony the King of Bola Bola sent his Comp^{ts} and a present of hogs and Fowls to the King of the ship, sending word also that he would in person wait upon him today. We therefore all hands staid at home in hopes of the honour of his excellencys visit. We were disapointed in our expectations not disagreably for instead of his majesty came 3 hansome lively girls who staid with us the morning and took off all regret for the want of his majesties company." They made a return visit to the King that evening "to see the great king and thank him for his civilities particularly of this morning." The King though proved as disappointing as the nature: "The King of the *Tata toas* or Club men who have conquerd this and are the terror of all other Islands we expected to see young lively hansome &c &c. but how were we disapointed when we were led to an old decrepid half blind man who seemd to have scarce reason enough left to send hogs, much less galantry enough to send ladies."

While based around Raiatea and Huahine Solander took the opportunity to visit Tahaa ('Otaha') briefly while the long-boat's crew were trading for more supplies.

On 9th August they "again Launchd out into the Ocean in search of what chance and Tupia might direct us to." Banks was not sorry to leave Polynesia, for he longed for the great discoveries that awaited in the great undiscovered continent – Terra Australis. He was to be frustrated though as each new discovery turned out to be something other than sought after land: first New Zealand was realised to be a group of islands and then Australia, although a continent, was not on the scale anticipated.

Despite the disappointment of missing Terra Australis, the expedition was a great success, in that the Transit of Venus had been observed and measured accurately and Banks had collected many new plants and animals. Back in London, Solander moved into Banks's house as his secretary and librarian, where he remained until his death in 1782. In 1773 he was made keeper of the Natural History department at the British Museum.

Terra Australis remained an enigma and a strategic prize, and so Cook was again commissioned to explore the southern oceans in 1772. Banks was also commissioned as Naturalist and the 'Resolution' fitted out to meet his requirements. The ship's waist was heightened, an additional upper deck fitted and the poop deck raised, at the great cost of £10,080 12s 9d. When it came to sea trials however, the ship turned out to be top-heavy and the extra decks had to be removed, bringing the cost up to £10,962 15s 9d. Banks considered these "adverse conditions" to be unacceptable and refused to travel. He was replaced by the German pastor and amateur ornithologist Johann Reinhold Forster and his son Georg. Johann Forster was 43 at this time and his son just 17. Because Georg was so young, it was not thought necessary to make him sign the usual documents which prevented the ship's crew from publishing any account of the voyage until after the official reports had been published. Georg took advantage of this lapse and published *Observations Made during a Voyage round the World* in 1778, three years after their return. This caused a scandal at the time, but does leave a rare detailed account from the naturalist's perspective and contributed to his election to the Royal Society at the age of 22.

The 'Resolution' and 'Adventure' set out on 13th July 1772, sailing round the Cape of Good Hope, across the Indian Ocean and into the Pacific. At Cape Town Forster took on an assistant, Anders Sparman. Sparman was a Swede, and another pupil of Linnaeus.

Although it had been visited by only three European vessels at this time, Tahiti was already a longed for destination to the explorers. According to Georg Forster, at sunset on 15th August 1773 the crews of the 'Resolution' and 'Adventure' "plainly saw the mountains of that desirable island, lying before us, half emerging from the gilded clouds on the horizon. Every man on board, except one or two who were not able to walk, hastened eagerly to the forecastle to feast their eyes on an object, of which they were taught to form the highest expectations, both in respect of the abundance of refreshments, and of the kind and generous temper of the natives, whose character has pleased all the navigators who have visited them."

Their expectations were met when the sun rose the following day: "It was one of those beautiful mornings which the poets of all nations have attempted to describe, when we saw the isle of O-Taheite, within two miles before us. The east-wind which had carried us so far, was entirely vanished, and a faint breeze only wafted a delicious perfume from the land, and curled the surface of the sea. The mountains, clothed with forests, rose majestic in various spiry forms, on which we already perceived the light of the rising sun: nearer to the eye a lower range of hills, easier of ascent, appeared, wooded like the former, and coloured with several pleasing hues of green, soberly mixed with autumnal browns. At their foot lay the plain, crowned with its fertile bread-fruit trees, over which rose innumerable palms, the princes of the grove. Here every thing seemed as yet asleep, the morning scarce dawned, and a peaceful shade still rested on the landscape. We discerned however, a number of houses among the trees, and many canoes hauled up along the sandy beaches."

As soon as the inhabitants of those houses woke the explorers received the famed Polynesian welcome. Although Georg later became more interested in the people than the land, his initial reactions were to start what he had travelled so far to do. "I immediately began to trade for natural productions through the cabin-windows, and in half an hour had got together two or three species of unknown birds, and a great number of new fishes, whose colours while alive were exquisitely beautiful. I therefore employed the morning in sketching their outlines, and laying on the vivid hues, before they disappeared in the dying objects."

One of the early realisations that Georg made was that everyone had misunderstood the name of the place. Wallis had not bothered with the island's true name and had named it himself: King George III island. Cook had taken more trouble and had understood the island to be Otahiti, which was later to give rise to the snail name *Partula otaheitiana*. Georg was more interested in the people and took the trouble to learn the language of the Polynesians. He "immediately found that the O or E with which the greatest part of the names and words in lieutenant Cook's first voyage, begin, is nothing else than the article, which many eastern languages affix to the greater part of their substantives. In consequence of this remark, I shall always in the sequel either omit this prefix, or separate it from the word itself by a hyphen". Bougainville had also understood correctly and as Georg noted he "expressed it as well as the nature of the French language will permit, by Taïti, which, with the addition of a slight aspirate, we pronounce Taheâtee or Tahitee." The spelling of the name was eventually to settle as Tahiti.

The Forsters and Sparman obtained much of their material from the Polynesians. Some of this was of limited value, but sometimes they procured worthwhile specimens: "[Tuesday 17.] Seeing that I enquired for plants, and other natural curiosities, they brought off several, though sometimes only the leaves without the flowers, and vice versa; however, among them we saw the common species of black night-shade, and a beautiful *erythrina*, or coral-flower; I also collected by these means many shells, coralines, birds, &c."

The interior of Tahiti presented an awesome sight: "[Wednesd. 18.] ... it ran up between the hills into a long narrow valley, rich in plantations, interspersed with

Fig. 5. Johann and Georg Fortser on Tahiti, 1780. Painting by J.F. Rigaud, original in National Portrait Gallery, Australia.

the houses of the natives. The slopes of the hills, covered with woods, crossed each other on both sides, variously tinted according to their distances; and beyond them, over the cleft of the valley, we saw the interior mountains shattered into various peaks and spires, among which was one remarkable pinnacle, whose summit was frightfully bent to one side, and seemed to threaten its downfall every moment." Nothing daunted though the naturalists started work: "Our first care was to leave the dry sandy beach, which could afford us no discoveries in our science, and to examine the plantations, which from the ships had an enchanting appearance, notwithstanding the brownish cast which the time of the year had given. We found them indeed to answer the expectations we had formed of a country described as an elysium by M. de Bougainville. We entered a grove of bread-trees, on most of which we saw no fruit at this season of winter, and followed a neat but narrow path, which led to different habitations, half hid under various bushes. Tall coco-palms nodded to each other, and rose over the rest of the trees; the bananas displayed their beautiful large leaves, and now and then one of them still appeared loaded with its clustering fruit. A sort of shady trees, covered with a dark-green foliage, bore golden apples, which resembled the anana in juiciness and flavour. Betwixt these the intermediate space was filled with young mulberry-trees (*morus papyrifera*), of which the bark is employed by the natives in the manufacture of their cloth; with several species of arum or eddies, with yams, sugar-canes, and other useful plants."

Exploring further inland on the 19th "We came to the foot of the first hills, where we left the huts and plantations of the natives behind us, and ascended on a beaten path, passing through an uncultivated shrubbery mixed with several tall timber-trees. Here we searched the most intricate parts, and found several plants and birds hitherto unknown to natural historians."

"[Saturday 21.] ...we continued our walk, but turned towards the hills, notwithstanding the importunities of the natives, who urged us to continue on the plain, which we easily perceived arose merely from their dislike to fatigue. We were not to be diverted from our purpose; but leaving behind us almost the whole croud, we entered, with a few guides, a chasm between two hills. There we found several wild plants which were new to us, and saw a number of little swallows flying over a fine brook, which rolled impetuously along. We walked up along its banks to a perpendicular rock, fringed with various tufted shrubberies, from whence it fell in a crystalline column, and was collected at the bottom into a smooth limpid pond, surrounded with many species of odoriferous flowers. This spot, where we had a prospect of the plain below us, and of the sea beyond it, was one of the most beautiful I had ever seen, and could not fail of bringing to remembrance the most fanciful descriptions of poets, which it eclipsed in beauty. In the shade of trees, whose branches hung over the water, we enjoyed a pleasant gale, which softened the heat of the day, and amidst the solemn uniform noise of the waterfall, which was but seldom interrupted by the whistling of birds, we sat down to describe our new acquisitions before they withered."

However, although the initial impression was favourable they soon changed their minds: "Our three days excursions had supplied us only with a small number of species, which in an island so flourishing as Taheitee, gave a convincing proof of its high cultivation; for a few individual plants occupied that space, which in a country entirely left to itself, would have teemed with several hundred different kinds in wild disorder. The small size of the island, together with its vast distance from either the eastern or western continent, did not admit of a great variety of animals. We saw no other species of quadrupeds than hogs, and dogs which were domestic, and incredible numbers of rats, which the natives suffered to run about at pleasure, without ever trying to destroy them. We found however a tolerable number of birds, and when the natives gave themselves the trouble to fish, we commonly purchased a considerable variety of species, as this class of creatures can easily roam from one part of the ocean to the other, and particularly in the torrid zone, where certain sorts are general all round the world." Sadly, but not surprisingly, he did not comment on the snails!

An expedition of this sort would normally require many days sorting for each day of collecting but "Our acquisitions in natural history being hitherto so inconsiderable, we had leisure every day to ramble in the country in search of others, as well as to pick up various circumstances which might serve to throw a light on the character, manners, and present state of the inhabitants." Like Banks before him, Georg Forster was fascinated by the people. It was hardly surprising that the crew found the women a distraction as "the view of several of these nymphs swimming nimbly all round the sloop, such as nature had formed them, was perhaps more than sufficient entirely to subvert the little reason which a mariner might have left to govern his passions."

Their initial explorations were around the south-east, on the peninsula island of Tahiti-Iti. Later they sailed around the north. "[Wednesday. 25.] We were becalmed in the evening, and during a great part of the night, but had a S.E. wind the next morning, so that we stood in shore again, in sight of the northern-most part of O-Taheitee and of the adjacent isle of Eimeo [Moorea]. The mountains here formed larger masses, which had a more grand effect than at Aitepeha. The slopes of the lower hills were likewise more considerable, though almost entirely destitute of trees or verdure; and the ambient border of level land, was much more extensive hereabouts, and seemed in some places to be above a mile broad… In the mean while we gradually approached the shore, a faint breeze helping us on, and the evening-sun illuminating the landscape with the richest golden tints. We now discerned that long projecting point, which from the observation made upon it, had been named Point Venus, and easily agreed, that this was by far the most beautiful part of the island. The district of Matavaï, which now opened to our view, exhibited a plain of such an extent as we had not expected, and the valley which we traced running up between the mountains, was itself a very spacious grove, compared to the little narrow glens in Tiarraboo."

This area proved very different from their earlier explorations: "[1773. Sunday 29th August] On the 29th at day-break we landed at our tents, and proceeded into the country with an intention to examine its productions. A copious dew, which had fallen during night, had refreshed the whole vegetable creation, and contributed, together with the early hour of the morning, to make our walk extremely pleasant. ... The first part of our march was a little difficult, on account of a hill on which we mounted, in hopes of meeting with something to reward our trouble. But, contrary to our expectations, we found it entirely destitute of plants, two dwarfish shrubs, and a species of dry fern excepted. Here, however, we were much surprised to see a large flock of wild ducks rising before us, from a spot which was perfectly dry and barren, without our being able to imagine what had brought them thither from the reeds and marshy banks of the river, where they commonly resided. We soon crossed another hill, where all the ferns and bushes having lately been burnt, blackened our clothes as we passed through them. From thence we descended into a fertile valley, where a fine rivulet, which we were obliged to cross several times, ran towards the sea. ... On the sides of the hills we gathered several new plants, sometimes at the risk of breaking our necks, on account of the pieces of rock which rolled away under our feet."

Here they did manage to penetrate some way into the interior of the island: "[Monday 30.] Dr. Sparrman went on shore with me ... with a view to make another excursion into the interior parts of the country... From hence we proceeded up the valley, which having no rivulet in its middle, began to rise in proportion as we advanced. We resolved therefore to go upon the steep hill on our left, and with much difficulty accomplished our plan. ... At last we reached the ridge of the hill, where a fine breeze greatly refreshed us, after our fatiguing ascent. When we had walked upwards along that ridge for some time, exposed to the burning rage of the sun, reverberated from all parts of the barren soil, we sat down under the scanty shade of a solitary *pandang*, or palm-nut tree... Having rested a little while, we advanced up towards the interior mountains, which now appeared distinctly before us. The rich groves which crowned their summits, and filled the vallies between them, invited us to advance, and promised to reward our perseverance with a load of new productions. But we soon perceived a number of barren hills and vallies which lay between us and those desirable forests, and found it was in vain to attempt to reach them this day. We consulted amongst ourselves, whether we should venture to pass a night on these hills, but this was unadviseable, on account of the uncertainty of the time when our ships were to sail, and likewise impracticable for want of provisions.... We began to descent therefore, but found it more dangerous than when we came up: we stumbled every moment, and in many places were obliged to slide down on our backs."

From Tahiti they sailed to Huahine but the paradise of Tahiti had started to take its toll: "[Thursday 2.] ... A number of our people now felt the effects of their intercourse with the women at Matavaï Bay, and had symptoms of a disagreeable complaint. All the patients, however, without exception, had this disease only in a very slight and benign degree." From Huahine they sailed to Raiatea and here Georg again corrected Cook's island names: "[Wednesday 8.] ... which all the natives of Taheitee, and the Society Isles call O-Raietea, but which (upon what foundation I know not) is named Ulietea in captain Cook's charts." While based on Raiatea Georg joined a small party exploring Tahaa.

Although they seemed to have taken little trouble at Huahine, Raiatea prompted more exploration up one of the hills where they "found several new plants in the vallies, between them" and "in various rambles along the shores, in which we found many deep creeks towards the northern part, with marshes at the bottom, where wild-ducks and snipes resided in great plenty. These birds were more shy than we expected, which we soon learnt was owing to their being much pursued by the natives, who looked upon them as dainty bits."

After three days there they sailed on, exploring further afield in the Pacific. They did return to Tahiti some months later though, by then desperate for dry land, provisions and a kind welcome: "[Thursday 21.]We discovered land about ten o'clock the next morning, which in a few hours afterwards we knew to be part of Taheitee. We stood towards it all the day, but could not reach it before it became dark, and were therefore obliged to stay out another night. Every person on board gazed continually at this queen of tropical islands; and though I was extremely ill of my bilious disorder, I crawled on deck, and fixed my eyes with great eagerness upon it, as upon a place where I hoped my pains would ease. Early in the morning I awoke, and was as much surprised at the beauty of the prospect, as if I had never beheld it before. It was indeed infinitely more beautiful at present, than it had been eight months ago, owing to the difference of the season. The forests on the mountains were all clad in fresh foliage, and gloried in many variegated hues; and even the lower hills were not entirely destitute of pleasing sports, and covered with herbage. But the plains, above all, shone forth in the greatest luxuriance of colours, the brightest tints of verdure being profusely lavished upon their fertile groves; in short, the whole called to our mind the description of Calypso's enchanted island."

Georg was too ill to explore and remained in his cabin for several days, but did purchase specimens from the canoes outside the window. In the meantime Dr. Sparrman and Johan Forster tried to climb the mountains. On 29th September they returned, having found the climb extremely difficult:

> "The difficulties increased as they ascended; the paths running along the narrow ridge of steep hills, whose sides were almost perpendicular. The greatest danger arose from the slipperiness, occasioned by the rains of the preceding day. When they had ascended to a considerable height, they found thick shrubberies and woods on these steep sides; and attempting to collect plants there, they frequently met with precipices which are really tremendous.

Still higher up the whole ridge was covered with a forest, where they gathered a number of plants, which they had never seen in the vallies below. After they had crossed the ridge, there fell a heavy shower of rain; and they were coming to a very dangerous part, Tahea said they could go no farther. They resolved however to leave their heavy plant and provision bags behind, and went up with a single musket to the summit of the mountain, which they reached in about half an hour. Just about that time the clouds broke, and they had a prospect of Huahine, Tethuroa, and Tabbuamanoo. The view of the fertile plain under their feet, and of the valley of Matavaï, where the river makes innumerable meanders, was delightful in the highest degree. Thick clouds however prevented their discerning any thing on the south side of the island. In a few moments even the other part was covered again, and they were involved in a mist which wetted them to the skin. In their descent my father had the misfortune to fall in a very rocky place, and bruised his leg in such a manner, that he nearly fainted away. When he recovered, and attempted to proceed, he found that he had also received a dangerous rupture, for which he now continues to wear a bandage. Tahea assisted him in going down; and they all arrived on board about four o'clock in the afternoon. The upper hills they found to consist of a kind of clay extremely compact and stiff. The vegetation on the upper part of the mountains was luxuriant, and the woods consisted of many unknown sorts of plants."

On 15th May 1774 they were back at Huahine, this time exploring for several days. By 24th May they were on Raiatea, on the 28th exploring the east coast:

"The next morning we travelled along the shore to the southward, and met with a very fertile country, and hospitable people. From thence we walked on several miles, till we arrived at a spacious bay, where three little islands lie within the reef [Faaroa Bay]. The country round this bay was swampy, and well stocked with ducks. Here we passed some time in shooting, and then embarked in two small canoes, and were safely landed at one of the little islands. We found a few coco palms and shrubs, but no fruit-trees upon it; and there was only a single fisherman's hut, containing some nets, and other fishing-tackle. We returned very soon to the main shore, having found no shells, though the hope of meeting with some had principally induced us to cross the water."

On their final departure from Polynesia Georg wrote "Thus we left an amiable nation, who, with all their imperfections, are perhaps more innocent and pure of heart, than those who are more refined and better instructed. Without quoting the example of Mahine, we have often been witnesses to reciprocal acts of kindness, which convinced us, that the social virtues are frequently exercised amongst themselves. I have seen a single bread-fruit, or a few coco-nuts shared between a number of people, so that every one partook of them."

This view of the Polynesian's egalitarianism fitted into Georg's outlook on the world very neatly. Whether it influenced him in any way, or whether he interpreted the islanders through the prism of his existing opinions is not known. Either way he later became something of a revolutionary. In Germany he established a Jacobin club and this perceived collaboration with the French Republic led to the German Emperor declaring him an outlaw and placing a prize of 100 ducats on his head. Banished from his native country and isolated from his wife and children, he remained in France. Revolutionary France was not the equitable paradise he had dreamed of, but he remained true to the revolution, seeing the Reign of Terror as merely a necessary stage. In January 1794 at just 39 he died of a stroke.

Cook's second expedition sailed further south than any previous voyage and proved that Terra Australis did not exist. As such it was a scientific success. Did it collect *Partula*? That remains unknown. Georg Forster seems to have taken more interest in non-botanical specimens than did Banks, but he still failed to record exactly what he collected.

The second expedition led on to a third in 1776. The purported aim of this was to return Omai, a Raiatean who had travelled with the second expedition, home to Polynesia. In reality the Admiralty had a strategic aim, to discover the Northwest Passage. The ships 'Resolution' and 'Discovery' were equipped and William Anderson employed as surgeon-naturalist. Anderson was not a naturalist but had served as surgeon's mate on the second expedition. The voyage was ill fated, with Cook being killed in Hawaii and Anderson dying on the return journey. Very little is known of what he collected.

Chapter 2. The shell collectors

Partula lutea

Somewhere in the material collected on one of Cook's voyages by Banks, the Forsters or Anderson were a few *Partula* shells. One of these came into the possession of Thomas Martyn. He may have obtained Banks's or Forster's but he is known to have bought most of the sea- and land-snail shells from the third voyage. On 9th December 1780 he wrote: "I may venture to affirm that I have purchased, amounting to 400 guineas, more than 2 thirds of the whole brought home. Nevertheless I do not abound either in variety of the new or many duplicates of the known ones that are valuable" and on the 18th: "I repeat it again 2/3rds. of the whole from the 2 ships were bought by me. All the rest of the shells are positively in the cabinets of Mr Banks, Dr Fordyce & two other Gentleman". Others ended up in the collections of the Duchess of Portland, the Countess of Bute and John Hunter, and probably others as well, but these three collections were used by Martyn in his '*Universal Conchologist*'.

Martyn resided in London from around 1781 to 1816. He was a natural philosopher who received renown for his artistry, including the receipt of a medal from Pope Pius IV and honours from kings of Europe. His interests were broad and he published widely, including an essay on ballooning, an edition of Harris's '*Natural System of Colors*' and political pamphlets on such things as 'Great Britain's Jubilee Monitor', a national assessment for the maintenance of disabled soldiers and sailors, and one railing against Napoleon Bonaparte. His most important work though was one of the greatest illustrated books on shells, the '*Universal Conchologist*', in 1784.

Martyn intended to produce a great work covering all the shells known at the time, starting with the great wealth of new species being collected by the expedition in the South Seas. He wished to improve on the then current practice in natural history publishing which he felt was inadequate because "complicated systems, bad arrangements, and the practice of crowding many sheets of different families into one plate, have not only confused the subject, and created a distaste to the science itself, but made it necessary that even the most experienced collector should have some clew to conduct him through those labyrinths of difficulties." His works were to have clear figures, with full detail and with an eye to display.

This was a costly exercise and he soon realised that he would be unable to commission professional artists for the task. Instead he recruited boys who could be trained to his requirements. These would produce the illustrations to a standard, uniform style, impossible from professionals. These boys were to be "born of good but humble parents" who "could not from their own means aspire to the cultivation of any liberal art, at the same time that they gave indications of natural talent for drawing and design." One such boy was identified and trained, and within a year was good enough to act as a tutor for two more. In two years this small workshop was capable of producing illustrations of such quality that orders for the planned work were secured. One year later seventy copies of the first two works (80 plates) had been completed. However, Martyn was not satisfied with these, feeling that they lacked the quality of the original test plates and rejected all of the plates. Rumour of this demanding exaction did no harm to the project's reputation. Eventually the copper etchings were produced to Martyn's satisfaction. From these were printed the illustrations which were then hand coloured. By this time Martyn had "sunk in it no inconsiderable share of a private competence" but despite the cost he was satisfied that it was "as worthy of himself, of his country, and of the learned world as art and his utmost abilities of every kind could effect." His satisfaction was not just with the illustrations, feeling that with the workshop itself (now numbering nine) he "would feel it a nobler boast to have educated one good citizen than any number of artists, however ingenious."

Martyn's book is not only a beautiful example of 18th century illustration but the most remarkable marriage of science and artistry ever produced. It is in this that *Partula* makes its first appearance, although ironically in one of the few plates that fails to meet the standard of excellence that Martyn sought. One cannot but suspect that he failed to give the plates of land snails the same attention as he did the more collectable sea shells. This would not be remarkable, for the collectors themselves brought back remarkably few specimens and recorded little of significance. Martyn's figure on plate 67 of the second volume is the *Partula faba* brought from Raiatea by one of Cook's expeditions. Martyn called this '*Limax faba*'. This name would appear to be the use of the Linnaean Latin binomial system. Martyn had used Linnaeus's system in his insect works, but the shell book was his special creation and for this he decided to use a system all of his own. Instead of long descriptions he would use perfect illustrations that would convey the essential characters of the shells and so "stand on the firm and unalterable basis of truth and nature." "The work will commence with the figures of the shells (most of them rare and nondescript [meaning new, undescribed species rather than uninteresting ones]) which have been collected by the several officers of the ships under the command of Captains Byron, Wallace, Cook, and others in the different voyages made to the South Seas." The species were to be named in a planned

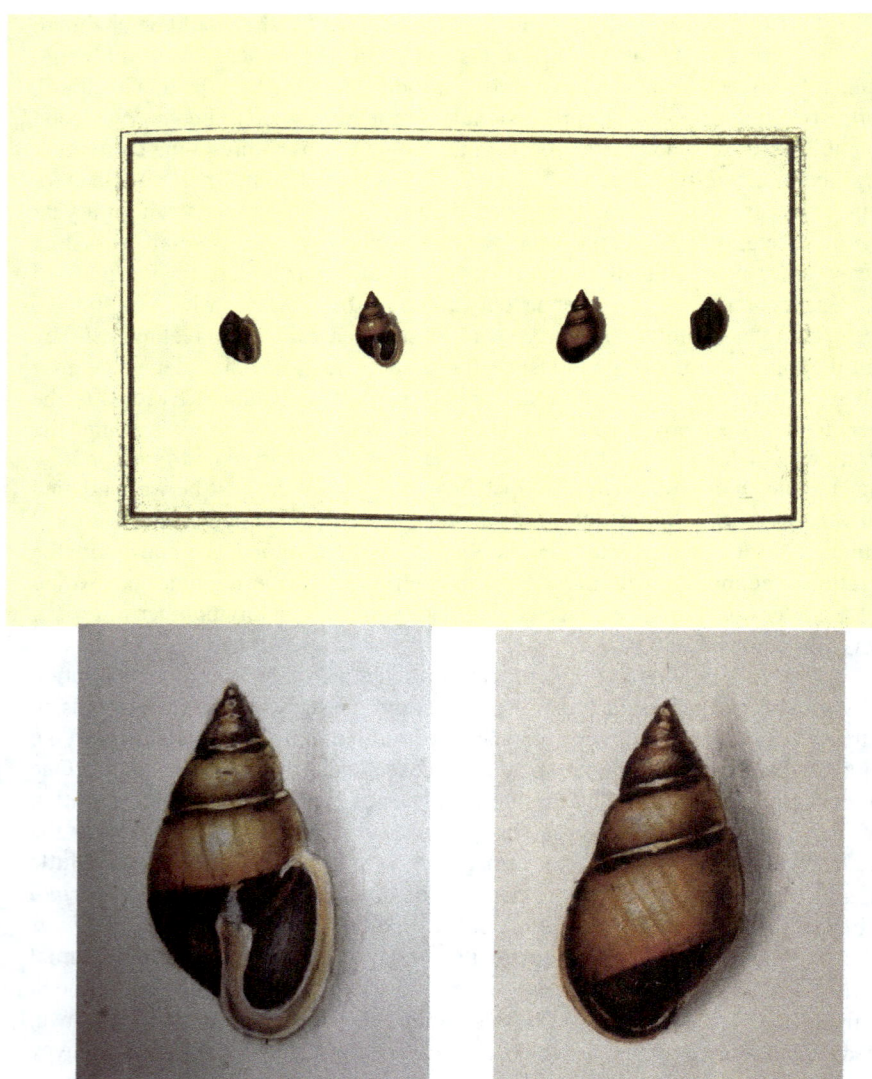

Fig. 1. Martyn's '*Limax faba*'. Top left – original plate. Top right – details of *Limax faba*. Bottom – detailed views. The original plates show Martyn's distinctively uncluttered plates and wide borders dating from a time when the appearance of the plate was the primary concern, and not the cost efficiency of printing.

synoptic table and an explanatory table with "the English name and family with an initial letter denoting the genus or division of the family to which the shell belongs, according to the system of the author; thirdly, the Latin name; fourthly, where the shell is found; and lastly, in what cabinet it is preserved."

Martyn bought much of the Cook material and the '*Limax faba*' was listed as being in his collection. There is no record of which shells he bought, but he had little opportunity to buy specimens from the first two voyages, these mostly being bought up by other collectors. The keenest purchaser of Cook's material was the London based dealer and collector George Humphrey. Humphrey had his own museum, the Museum Humfredianum, albeit for only one year (1778-9) as bankruptcy forced him to close his doors. In 1771 he had started a project to publish the first substantial British shell book, the '*Conchology, or natural history of shells*', to be written by Emanuel Mendes da Costa, a Clerk to the Royal Society. Unfortunately da Costa had embezzled money from the Royal Society and his attempt to complete the book while in prison was not a success. The book was never finished, giving Martyn's '*Universal Conchologist*' the honour of being the first great shell book published in English.

Humphrey is more notable as a dealer than as an author. Humphrey bought shells from the first voyage and many from the second voyage direct from the crew, spending around £150 on shells from the 'Resolution'. Only the third voyage escaped him as Martyn had bought up the bulk of that collection, leaving Humphrey only able to spend £20 on whatever the crew had collected. Some of Humphrey's acquisitions ended up in the collection of Margaret Cavendish Bentinck, Duchess of Portland. The Duchess was the richest woman in Britain, an assiduous collector of almost everything. She supported some collectors in their excursions, including providing some financial support to Captain Cook's expeditions. Aristocratic collectors and patrons were not unusual in the 18th century but the Duchess of Portland's collection was different. The Duchess took the intellectual aspect of collections of curiosities seriously; this is not surprising as she was a member of the 'Society of Blue Stockings' – an informal association of women dedicated to discussion of intellectual matters, principally the arts. This was a conscious movement away from the superficial pursuits that dominated most aristocratic gatherings in Britain and are so conspicuous in literature of the time.

During her lifetime the collection became famous as it, along with its associated zoo and gardens, was open to the public. It was famed for its size, probably being the largest collection of objects ever amassed by a single individual. It was also notable for the academic industry that attended it, being known as 'The Hive'. Unlike most collections this was well-curated, by the resident botanist, the Reverend John Lightfoot, and latterly by Daniel Solander. The Duchess aimed to have the collection organised scientifically and Solander was working to this end, cataloguing her collection, when he died.

The Duchess did acquire some *Partula*, for Lightfoot's catalogue of her collection (it was auctioned after her death, taking 39 days to work through the 4,156 lots) includes 'one *reverse snail from Otaheite*'. As the bulk of her collection was obtained by the British Museum these should be among the specimens in the Natural History Museum, London. Unfortunately most of the old specimens have lost their labels and it is impossible to identify any with certainty as being from this great collection.

Fig. 2. Margaret Cavendish Bentick, Duchess of Portland, portrait after M. Dahl, date unknown. Original at Hardwick Hall, National Trust.

For the '*Limax faba*' specimen, the trail seems to go cold. It is not known which voyage Martyn obtained it from, nor what happened to it: his collection vanishes. All we can say for certain is that it was not a shell collected by Banks, as Banks failed to open the box of shells he collected until 1787, after the date of the '*Limax faba*' illustration. There is no evidence that Martyn maintained his collection of shells after the publication of the '*Universal Conchologist*'. After publishing his great work on shells he moved on to books on beetles (1792), spiders (1793), plants (1795) and finally moths and butterflies (1797). None of his later works deals with biological topics at all. Given that his shell collection seems to have served its purpose he may have sold it before his death in 1816. Of all the dealers alive at the time, Humphrey stands out as the most acquisitive and it is likely that he would have obtained at least some of Martyn's shells, either in the 1790s or after the illustrator's death.

Humphrey leads to another thread in this story. He sold many specimens to the Danish artist and naturalist Lorenz Spengler, and this may have been the source of a number of *Partula* shells in Spengler's collection. Spengler was the Keeper of the Royal Cabinet of Curiosities to the Danish King and a compulsive collector. Such was the scale of his acquisitions that he was forced to move house to accommodate his ever expanding collection. His collecting habit was funded by his craftsmanship, being a cabinet maker and, more importantly, an amber and ivory turner. The latter meant that he was highly regarded as a manufacturer of false teeth, a lucrative trade. Spengler's specimens formed a large part of the shells illustrated by Johann Hieronymus Chemnitz in the '*Neues systematisches Conchylien-Cabinet*'. This series of volumes had been started by Friedrich Wilhelm Martini and continued by Chemnitz after Martini's death in 1778. The 1795 volume included a large portion of Cook's specimens, based on the shells in Chemnitz's collection, and also Spengler's. In these is 'Auris mida fasciata', clearly recognisable as *Partula faba*. The figure is not up to Martyn's standard but the species is clearly the same.

The *Partula faba* on Plate 121 is described as "Das bandirte Midaohr aus ben sudlandern - Auris Midae fasciata terrae australis." Chemnitz noted "This type of Midas-ear came from D. Solander. He collected it from the southern-lands." (original in German, free translation). Finally we have the collector - it was Solander who collected this specimen, on the first of Cook's voyages.

A second species of *Partula* appears now, for Chemnitz illustrated *Partula otaheitana* on plate 112. He described it as "Die linte Otaheitische Flusschnecte - Helix perversa, in rivulis insulae australis Otaheite reperta." This was certainly one of Cook's specimens for, as Chemnitz specified: "These strange snails come from southern lands, obtained from the Cook voyages. They came from England with the following description: Small reverse long Snail found in the rivers of Otaheite" (free translation). Although *faba* specimens from the Cook expeditions are mentioned as single specimens it seems that *otaheitana* was more abundant, for elsewhere in the '*Conchylien-Cabinet*' Chemnitz writes: "It is from the Cooke [sic] expeditions to the reasonably well known Island of Otaheite, in the South Seas, found in small creeks and rivers. I obtained two examples of the same from London, under the name "reverse long river snail from Otahite"". He also noted that they were not uncommon and were to be obtained cheaply, indicating that many specimens had been brought back by the crew of the expedition.

Fig. 3. Lorenz Spengler and the *Partula* shells in his collection. Portrait of Spengler by C.G. Pilo, 1751, in the Museum of National History at Frederiksborg Castle

Although Chemnitz recorded his *faba* and *otaheitana* as "Ex Museo nostro" (from our museum) rather than "Ex Museo Spengleriano", neither species appears in the catalogue of the auction of his collection following his death in 1803. This might indicate that they were actually in Spengler's collection. Spengler's specimens were purchased by the Danish Royal Natural History Museum. On the death of the King 1848 they were transferred to the Zoology Museum of the University of Copenhagen where several Spengler *faba* and *otaheitana* specimens are preserved. Chemnitz's own collection was auctioned and bought (possibly in its entirety) by the dealer Cetti. He sold it on to the Academy of Sciences of the Russian Empire. This is now in St. Petersburg, although it has lost its original documentation and only a small part of the collection has been recognised. In theory the Chemnitz *Partula* should have ended up in St. Petersburg, but it seems that some specimens were transferred to the Royal collection along with Spengler's shells. One of the Copenhagen Museum's *faba* specimens was labelled as being from the Chemnitz collection. The label is by Hendrik Beck, Prince Christian Frederick's (later King Christian VIII of Denmark) shell expert. When Beck catalogued the Prince's collection in 1837 he listed ten *Partula* species, including several varieties of *faba* and *otaheitana*, and specified the origin as 'Ulitaea' (Raiatea). This seems to have been an original piece of information, associated with one of the shells as this is the first published locality. Specimens with good locality data had been collected nine years earlier, but that information would not be published for a further 14 years.

The shell in the Copenhagen Museum labelled as 'Chemnitz?' bears a very close resemblance to Chemnitz's figure. Unfortunately, it lacks any other details. The Copenhagen collection includes three shells with labels in Spengler's handwriting: one *faba* and two small shells numbered 67 (confusingly the same number as Martyn's *faba* illustration). These are catalogued as *otaheitana* but are in fact a third species, *affinis*. It seems clear now that several *Partula* were collected on Cook's expeditions: *faba*, *otaheitana* and *affinis*. At least one of the former was from the first expedition, but the whereabouts of the very first *Partula* to be recorded (Martyn's *Limax faba*) remains a mystery.

When Martyn devised his system of organising shells he often disregarded Linnaeus's names, leading to some confusion and he was also unfortunately inconsistent. There was no fixed species naming system at the time and Martyn's approach was idiosyncratic. We now employ Linnaeus's binomial system of two names: a genus (group) name, followed by the species name. This is laid down as a rule by the organisation that regulates the modern system, the International Commission on Zoological Nomenclature. The Commission dictates that all names species must have such binomial names formed in Latin, or at least

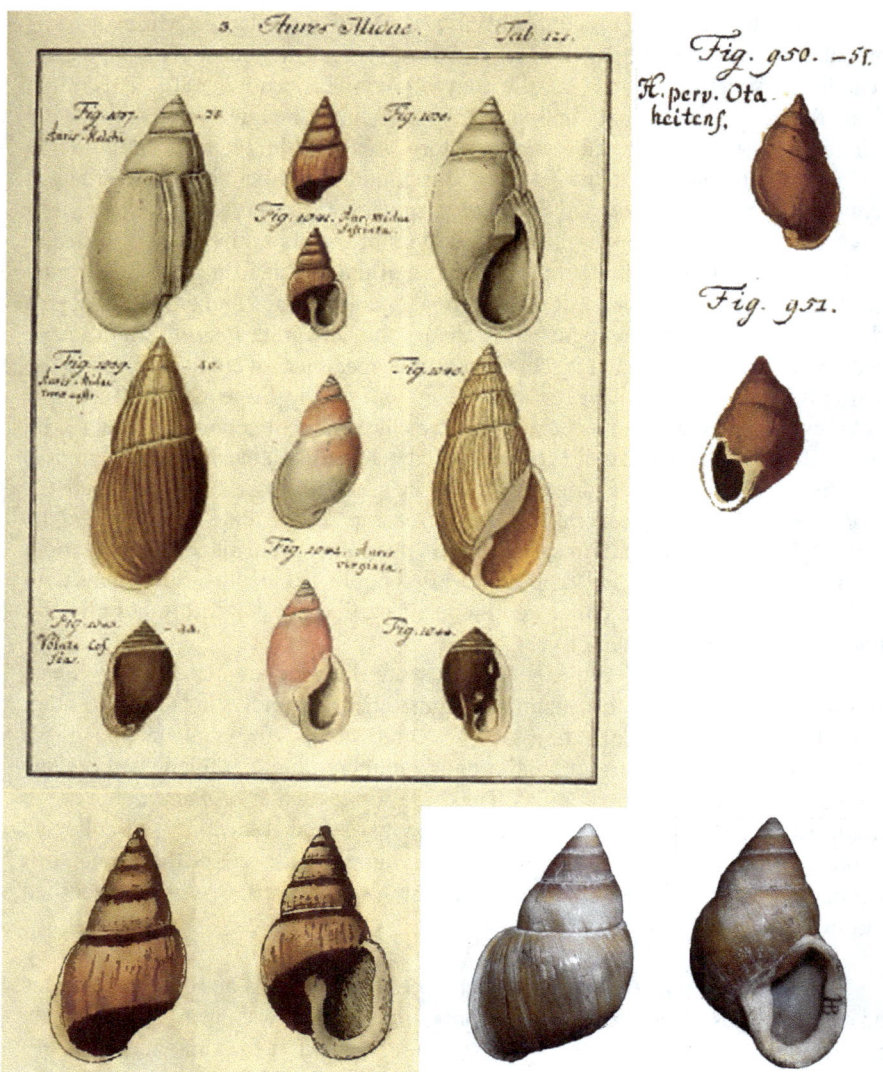

Fig. 4. Chemnitz's 'bandite Midaohr' and 'Helix perversa'. Top left – the original plate showing the 'bandite Midaohr'. Top right – 'Helix perversa', edited for position. Bottom – detail of the 'bandite Midaohr' compared with the possible Chemnitz specimen in the Copenhagen Museum. The full plate shows the more standard plate design used in the '*Conchylien-Cabinet*'

Latinized, and written in a font that is distinct from the rest of the text – usually taken to be in italics. In this book I have compromised the nomenclature slightly, in usually missing out the genus name to avoid the cumbersome repetitive use of '*Partula*'.

As Martyn did not use the binomial system throughout his book the Commission later declared most of Martyn's names unavailable for the purposes of taxonomic nomenclature. Martyn was mostly binomial, but in some cases omitted the genus, thereby becoming mononomial, and in some cases trinomial. This was also the case with Linnaeus! The result of this is that Martyn's *Limax faba* is nothing more than a beautiful curiosity. Similarly, and more obviously, Chemnitz's '*Auris Midae fasciata*' and '*Helix perversa*' are also deemed invalid.

Although the names coined by Martyn and Chemnitz were ruled inadmissible their efforts were not completely overlooked. The value of Martyn's work was recognised by Johann Friedrich Gmelin who incorporated the species in '*Universal Conchologist*' into the Linnaean scheme in 1791 in his edition (the 13th) of Linnaeus's 'Systema Naturae'. *Limax faba* was placed under the genus *Helix* as "Faba. 252. H. testa imperforata laevi crocea, anfractuum margine basique sulcis, aperture caeruleu. *Mart. univ. Conch.* 2. *t.* 67. *med. Habitat in* Tahiti." This statement of origin repeated Martyn's record from his explanatory table of 'Otaheite', presumably using Tahiti as a general area rather than a specific island. By this action Gmelin's *Helix faba* becomes the first valid taxonomic name for this species. Gmelin's understanding of snail taxonomy had moved on from Martyn's, he recognised that snails and slugs were a diverse group and while Martyn used *Limax* for most snails, Linnaeus (and hence Gmelin) used it explicitly for slugs. When Gmelin incorporated Martyn's species into the Linnaean system he moved '*Limax*' *faba* into a snail genus '*Helix*'. However, this was only a slight improvement as *Helix* was Linnaeus's term for any snail.

'*Helix perversa*' was less fortunate. It simply meant the perverse snail; that is, the snail that coiled the wrong way - with a left-handed or sinistral spiral, rather than the usual right-handed or dextral one. This feature of left-handed spirals was to prove to be remarkably interesting nearly 200 years later, but at this time was just a curiosity. It was noted for several different species, so '*Helix perversa*' never applied to one species and Gmelin did not pick up Chemnitz in the way he did Martyn. However, Chemnitz's *Helix perversa* would remain without a valid name for only one more year. In 1792 the French malacologist Jean Guillaume Bruguière named it *Bulimus Otaheitanus*. Unlike Gmelin, Bruguière was working with actual shells, and did not just repeat the description from Chemnitz, although he did copy the supposed locality, reporting that this freshwater species came from the same voyage as the *Bulimus australis*, which is what Bruguière chose to rename the '*Helix' faba*. For both he specified that they had come from Cook's first voyage. Bruguière gave a very detailed description and noted that a tooth-like projection was present in the mouth of the shell in adults, but not juveniles. Clearly he had a range of specimens before him, probably obtained through Humphrey's dealership.

By this stage the classification of snails had moved on one more step; no longer were all land molluscs lumped into the *Limax* slugs and the *Helix* snails. Now a distinction was being made between the round *Helix* and a newer group of elongate shells – *Bulimus*. This still amalgamated a wide variety of unrelated snails but in 1822 the Polynesian tree snails were recognised as distinct. André Étienne d'Audebert de Férussac named them *Partula*, after the Roman goddess of childbirth, in reference to finding fully developed embryos inside them, suggesting that they had live birth. He gave a long description of the genus and then included several species within it. The first of these was '*Helix pudica* Mull.' Müller's *Helix pudica* is a confused species; based on Férussac's usage Anton later listed it as a synonym of *Partula faba* although the true Müller's *Helix pudica* is in fact the South American *Strophocheilus pudica*. The second of Férussac's *Partula* was *Partula australis*, Bruguière's new name for Martyn's species. He then gave more detail on *otaheitana* from "The island of Otaïti. It is clearly a mistake that it is said to be a stream species" (original text in French).

Following Cook's expeditions, two French voyages came back with *Partula*. In 1819 the 'Uranie' commanded by the explorer Louis Claude de Saulces de Freycinet arrived in the Marianas and Tahiti. In the Marianas islands the expedition's zoologists Jean-René Constant Quoy and Joseph Paul Gaimard collected two new *Partula*: *gibba* and *fragilis* (later to become *Samoana fragilis*). It would seem probable that they also encountered *otaheitana*, but they did not record this. The voyage of the 'Uranie' was a particularly well supported expedition, with the two zoologists and a botanist in the person of Charles Gaudichaud-Beaupré. Although Cook's naturalists had collected prolifically in the Pacific there was no shortage of new plant species for the young botanist to discover. Among these was a climbing *Pandanus* or screw-pine which he named *Freycinetia* after their leader. Although Gaudicaud-Beaupré does not seem to have commented on any snails he must have had to pick them off his specimens before pressing them, for a *Freycinetia* plant was the favourite home of many Polynesian *Partula* species.

Three years later Freycinet's hydrographer Louis Isidore Duperry was himself in command of a circumnavigation. His ship, the appropriately named 'La Coquille', called in at Tahiti and Bora Bora. Yet again, Tahiti yielded *otaheitana*. The naturalist René Primevère Lesson recorded: "This little shell is abundant under the leaves, in the cool places of Matavi valley, at Point Venus, on the island of O-Taiti, but particularly on the summit of the central mountain of Borabora" (original text in French). Lesson took an interest in the snails of the expedition and the descriptions seem to have been unusually accurate for the time. Lesson recognised that the Bora Bora shells were not the same as those of Tahiti and named them *lutea* in 1831. The comment that *otaheitana* was particularly common on Bora Bora remains odd given that no-one since then has found anything other than small round, dextral *lutea* on the island, nothing like the large, slender, sinistral *otaheitana*.

In 1825 HMS Blossom, under the command of Captain Frederick William Beechey passed through Tahiti and in their collections *faba* appears once again. This was confusingly noted as being "found abundantly at Tahiti" despite that fact that *faba* does not come from Tahiti. The shell does seem to be accurately identified; the drawing of the shell looks like *faba* and it was identified by the Natural History Museum of London's Keeper of Zoology, John Gray. Did Gray provide the wrong shell for illustration, or did Beechey bring back *faba* from Raiatea despite not having visited that island?

At this time all Polynesian *Partula* were being reported rather vaguely from Tahiti and no-one was taking much care over their true origin. The shells that were appearing in Europe represented very sparse pickings from these expeditions. There were others appearing in Europe, but of unknown origin. Some of these are significant specimens, such as the Moorean *taeniata* described by Mörch's in 1850. This was "purchased from a whale fisher" and was thought to have come from Fiji. In reality *taeniata* was the most widespread species on Moorea and was to be particularly interesting to geneticists over 100 years later.

It was not until 1828 that significant numbers of *Partula* species were recorded. This was the year of the first Pacific voyage of Hugh Cuming. Cuming (1791-1865), 'The Prince of Shell Collectors', was perhaps the most obsessive shell collector ever. Born to a family of modest means in Devon in 1804, at the age of 13 he was apprenticed to a sail maker. After 15 years he left this trade to seek adventure in South America. He seems to have found what he sought, for he lived in Chile for several years. There he met the British consul Mr. Nugent and the conchologist Lieutenant Frambly. The knowledge and contacts of these two men proved invaluable in his new trade of collecting shells and plants to sell to collectors back in England. By the time he left Chile in 1826 he had amassed a significant fortune. Part of this he spent in commissioning a ship, designed specifically for collecting. He sailed his 'Discoverer' across the Pacific, collecting by dredge, shore and land.

In April 1828 Cuming was in Polynesia, travelling around Tahiti, Huahine and Raiatea ('Riitea' as he wrote):

"The Botanical productions of this Island I had not an opportunity
of investigating from the tempestuous Weather, with heavy Rains.
Birds I did not see one differing from the other Islands. I procured
Two species of Bulimus and a Helix from the Hills rising above the
Town - Marine shells I collected a great variety and very abundant,

Fig. 5. Hugh Cuming, photographed about 1860

amongst them several Rare ones. The Outer Reef taking of the force of the Sea made the Inner Reefs accessible to collect with Safety, amongst the shells I found new to me where Eleven Mitres, Three Cyprea one of them the Golden Cowry, Three Cerithiums, Two Solariums, a Cone, a Strombus, a Bulla, a Nassa, a Murex, a Natica, an Ovula, Two Olives, Five Terebras, and several others, a Pecten that I found under stones about an Inch and a half long, White, Imbricated, and Gaping is a singular shell. Crabs are highly interesting of this Island."

This description illustrates that it was the sea shells that obsessed Cuming. He collected land snails with interest, but not with passion. Nevertheless, from Raiatea he had two *Partula* (his 'Bulimus').

His specimens of *Partula* were later named by William John Broderip and Lovell Augustus Reeve in England and Ludwig Pfeiffer in Germany who gave locality information based on the information from Cuming. For each specimen Cuming recorded the island, a great breakthrough in biological exploration. However, his respect for locality data only applied to his own specimens; when he purchased material from others he paid little regard to any data that might exist. The *Partula* in the Cuming collection though were all Cuming collected and their data was published (even though it is no longer associated with the specimens). In these is the first accurate locality for *Partula faba*: 'Ulitea' (in Reeve's 1851 '*Conchologica Systematica*'). In fact, from Cuming's description of his collecting on Raiatea we can say that it came from "the hills above the town" or Uturoa.

Reeve later wrote: "It is to the collecting of shells that Mr. Cuming has mainly directed his attention; and it is chiefly owing to the care with which he has noted the habits and geographical distribution of their molluscan inhabitants that the studies of the conchologist have come to possess an interest of a philosophic kind which was formerly unknown."

Despite the inclusion of some locality data (not always consistent or reliable), there were problems with the Cuming collection, as noted by the Keeper of Zoology at the British Museum, John Gray:

> "No one knew better than Mr. Cuming the value of a new name to his specimens, as shown by his enmity to any one who doubted the novelty of the species described. He would not allow me to see his collection for many years after his return from South America, because I had pointed out that some of the shells which Messrs. Sowerby and Broderip had described as new were well-known species, and well figured by Chemnitz. Indeed, I was not allowed to see any part of his collection until it was first offered to the

> British Museum for sale, during his illness about sixteen years ago... The system that Mr. Cuming adopted of selecting three specimens of each variety or species most alike tended to prevent the number of nominal or presumed species from being observed during a casual examination of the collection, as it excluded those specimens which showed the transition from one variety to another which occurs in any given species—more especially as the species were not arranged in the drawers so that the most allied or presumed species were near to each other, but, on the contrary, the two or more variations of the same species were often placed in distant parts of the series."

Over his three expeditions Cuming assembled the richest collection gathered by a single individual up to that time, including 30,000 shell species and varieties. In the Philippine Journal of Science he is quoted as stating: "The greatest object of my ambition is to place my collection in the British Museum that it may be accessible to all the scientific world and where it would afford to the public eye a striking example of what has been done by the personal industry and means of one man." Cuming's pre-eminence in the shell collecting world was summarised by Samuel Pickworth Woodward in 1861:

> "The prince of shell-merchants, of course (facile princeps), is Mr. Cuming, of 80, Gower Street, who can supply whole collections, and many costly varieties which no-one else could obtain."

James Cosmo Melvill described Cuming in his later years when he financed others to collect shells for him:

> "I remember him as a somewhat stout, rubicund, good-humoured looking old man, with scanty, white curly hair, dressed in black, with open waistcoat, and white-frilled shirt front."

According to Edgar Leopold Layard, in his house in Gower Street Cuming

> "had a long plank table on trestles running the entire length of the room with its three windows. Along this he would walk, with a basket, or box, full of shells in one hand, from which he selected such specimens as he intended to supply to the collection making up. Placing them on the table, he would dictate to the secretary, name, author's name, and locality. These the young man wrote on a slip of paper already prepared, and placed by the specimens, which were afterwards packed by him."

In 1846 Cuming was seriously ill and offered his collection to the British Museum, offering to sell it to the Trustees for £6,000. Richard Owen wrote in fulsome support but John Edward Gray was sceptical of the collection's scientific value. The offer was turned down. Cuming recovered but for the remaining 19 years of his life was never in full health. At the age of 75 he died, and once again his collection was offered to the British Museum. Again the offer price was £6,000 and this time the museum did buy the collection, now numbering 82,992 specimens.

It has been claimed that the boxes of specimens were open to the elements and, it being a windy day, when the collection was carried into the museum all of the labels were blown away or jumbled. There seems to be no evidence for this story and it is more likely that the specimens were transported safely into the museum and there the Keeper of Zoology John Gray's wife Maria glued all the specimens and their labels onto wooden tablets, as was standard practice at the time. Over time the glue became brittle and in many cases what labels there were have become detached. Now there is no useful information on the specimens, and the only details we have for them were those published by Broderip, Pfeiffer and Reeve.

Within 20 years of Cuming's death he was already being blamed for the confusion we have today. In 1884 the first serious student of *Partula*, Andrew Garrett, wrote: "There is a good deal of doubt (in the genus *Partula*) in consequence of the uncertainty of the true types, which in some cases cannot be relied on as represented in the British Museum." The 'types' referred to by Garrett are the 'type specimens' upon which the names are based. When a species is named it is based on a particular specimen or group of specimens. These are the name-bearing types. If there is any subsequent uncertainty over the identity of the species, such as it being realised that there are actually two very similar species involved, then examination of the types should clarify the situation. The name should be based on one particular specimen, the 'holotype' and supported by a few additional 'paratypes'. In the 19th century whole groups were used for naming and no individual specimens highlighted, thus for most *Partula* species there are several 'cotypes' or 'syntypes', sometimes as many as a hundred such specimens. Unfortunately these sometimes turn out to be of different species, leading to much confusion.

Garrett's complaint continued: "With regard to Cuming's localities (who collected in this group some forty-five years ago), it really seems as if he intentionally gave the wrong habitats of some of the shells collected by himself, or marked them 'hab.?' in order to mislead the monographers and make as many species as possible. He did not appear to care so much about shells in a scientific point of view as to possess the largest collection in the world. Nothing vexed him so much as having any one doubt the validity of any of his new species, which he selected, generally three of each, from his stock of duplicates. He was also in the habit of changing the author's types whenever he obtained what he supposed to be better specimens. No doubt in most cases he was right in replacing the same species. Still, they were not the author's types."

Despite the confusion and lack of information in the 1860s, *Partula* were just starting to emerge from being curiosities and collectors' items into the focus of scientific study. This was not in the great collections of Europe, but far away, in the USA and the Pacific.

Chapter 3. The conchologist adventurers

Partula arguta

The expansion of shipping in the Pacific Ocean through the 19th century resulted in many more collections of *Partula*. Most will have been of isolated shells, collected as curiosities, unlabelled and undocumented. A few collectors are of note though, and one in particular stands out for an unusual label preserved with a shell in the Natural History Museum, London. A *Partula radiolata* from Guam is accompanied by a folded paper which reads:

"Partula radiata, Pf.
Island of Guam
Captain Thomas Rossiter 1854
picked out of about 2 Bushell of Partula gibba, Lesson
By me and Miss Rossiter 1862 who was my Wife after"

Thomas Rossiter was an American-born captain employed by the French government's fisheries department, and became a naturalised French citizen in 1838. Two years earlier he had made his first South Pacific voyage and returned to the same waters on a two year whaling expedition in 1840. His success in this and other voyages, including that which touched Guam in 1854, enabled him to purchase his own ship, the 'Wave of Goole' in 1859. Later that year he embarked his wife, his two teenage sons George and Richard, and his daughter Sophia on his ship and sailed for a new life in Australia. The children grew up in Sydney and there the young Rossiter men developed a close friendship with John Brazier. Brazier was also the son of a whaling captain, and had accompanied his father on several voyages in the 1850s, indulging in his passion for shell collecting. Along with George and Richard Rossiter, Brazier collected shells in the lands around Sydney. Such adventurous field-collecting would not have seemed suitable for a young lady but Brazier found that Sophia Rossiter shared his interest, helping him to sort and identify the shells he had amassed from his own collections, as well as those given to him by others, including her father. In 1872 they were married, and in the 10 short years of their life together he named many shells after her.

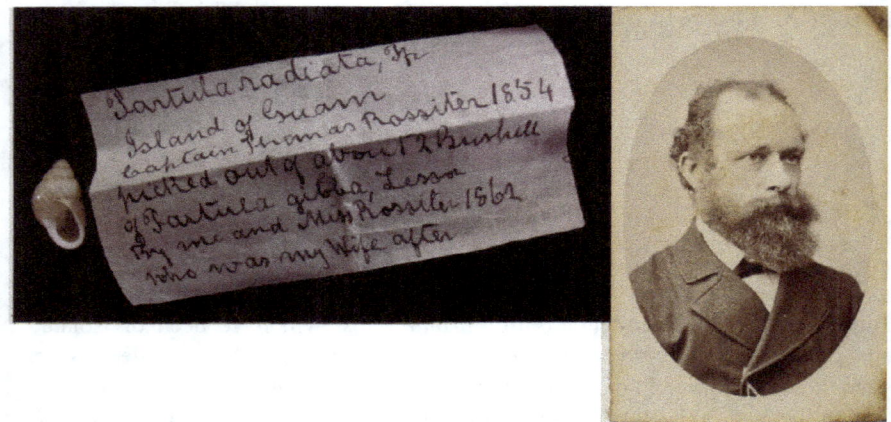

Fig. 1. The label recording John Brazier and Sophia Rossiter's sorting of Thomas Rossiter's Guam *Partula* specimens and photo of John Brazier (by Riisfeld & Co., 1883, in State Library of New South Wales).

Brazier sent many of his *Partula* to the conchologist William Harper Pease and later to William Dell Hartman. Unfortunately, although he was a prolific collector, he was still typical of the Victorian collectors – failing to record the origin of his shells accurately. Many of the new species in his collection were described by his friend, the doctor James Charles Cox. Cox exchanged shells with other collectors, including Andrew Garrett and William Harper Pease, who would both become central to the *Partula* story. In addition to the *radiolata* referred to above, a '*Partula bicolor*' was picked out of the same three bushels of *gibba* and named on a label (but never described) by Pease.

The appearance of William Harper Pease in the *Partula* story marks the end of the gentleman naturalists and professional explorers, and heralds the appearance of the adventurers who foreshadowed the scientists. In the Pacific islands several of these men were Americans who were drawn to the Kingdom of Hawaii. The islands were ruled by hereditary monarchs who had come to rely on the administrative abilities of expatriate Americans. These administrators were variously forward-looking businessmen and adventurers. Pease himself seems to have been something of a mixture of the two. Little is known of his early life. In 1848, at the age of 24, he is known to have been in Mexico with General Winfield Scott in the final days of the Mexican-American war. In what capacity is not known, but given his later life he may have been there in the role of some sort of surveyor, rather than in any direct military position. At the end of the following year he sailed from San Francisco to Hawaii, apparently speculatively, without having obtained any post in the islands prior to his arrival. In 1850 he became a citizen of the Kingdom of Hawaii and secured the post of Government Surveyor on the island of Kauai. One of his roles as surveyor was to manage the horse trade between the islands and from this he developed excellent contacts in the islands, including Haalelea, son of the Governor of Molokai and son-in-law to King Kamehameha I. It seems that Haalelea gave Pease one of his daughters, Nelly, as a gift; such provision of 'wives' was not uncommon in Hawaiian society at the time. By 1860, when Pease had moved to Honolulu, he had an office in Haalelea's house and kept his shell collection and library there.

In Honolulu Pease was appointed Assessor of the City of Honolulu, and Commissioner of Water Rights and Rights of Way. Despite these posts he was more interested in natural history and in shell collecting in particular: "That is all I think or care about" he wrote in 1865. The inheritance of his father-in-law's estate made him a wealthy man, enabling him to amass an extremely valuable library of rare books and facilitating the expansion of his extensive collection of Hawaiian shells. He was well known in high society in Honolulu, but according to Lady Franklin, widow of Sir John Franklin, the Arctic explorer, who met him at the plantation of Robert G. Wyllie, the Hawaiian King Kamehameha IV's minister of foreign affairs, he was not an appealing character. She wrote "I find it so difficult to make out what he says that much is lost to me — this proceeds partly perhaps from a want of some teeth in front of his mouth, but chiefly from his holding and chewing tobacco which not only thickens his speech, but causes him to be constantly spitting."

This unprepossessing-sounding man amassed a very important collection of Pacific land and sea-shells. Pease was not a simple collector though; he also took an academic interest in conchology and after the prolific Broderip, Pfeiffer and Reeve, Pease was the most significant describer of Cuming's Indo-Pacific shells. In 1857 his own collection had already grown to a notable size. Much of it he had obtained from other collectors, in the case of the *Partula*, from another adventurer, Andrew Garrett. In 1857 he wrote to Garrett:

"In land shells, especially *Achatinella*, my collection is complete, having all described species and about 20 undescribed, <u>distinct</u> [his underlining] species; not such as have been described by Pfeiffer and Newcomb, of which I am sure, one quarter are only varieties. Should you have any varieties on hand you may consider new, I would like to receive them, as I am about through describing those I have on hand."

Garrett was an American explorer, self-taught naturalist and artist. He was particularly interested in shells and fish. Many of his works are considered classics, including his multi-volumed catalogue of the *"Fische der Sundsee"*, and particularly his *"Terrestrial Mollusca Inhabiting the Society Islands."*

In 1839, at the age of 16 Garrett went to sea as a deck-hand and then as a crew member on whaling ships. In 1847 he reached Hawaii and was much taken with the place. At the conclusion of his whaling adventures in May 1851, he received a portion of the money raised by the sale of the sperm oil and had with him 20 packing cases of shells he had collected. At least some of these he sold to a shell store in Boston. In the following year he settled in Hawaii, staying there for the next seven years during which time he met Pease, with whom he shared his natural history interest.

Garrett's hobby of shell collecting stimulated him to take a more professional approach to natural history, learning Latin and becoming a competent self-taught artist. In 1855 he wrote to Louis Agassiz, professor at Harvard and founder of the Museum of Comparative Zoology, proposing that he should be commissioned to collect for the University. The Boston merchant and naturalist James M. Barnard was persuaded to donate the funds to pay Garrett to collect for eight years. Agassiz instructed him: "The principal merit of collections of objects of nature is not desired in our days from the accidental circumstances that they may contain new species but from the opportunity they afford of elucidating natural laws. The collector ought, therefore to have his attention constantly turned to this important end and must on that account collect in a particular way…" This he did, in Hawaii and in other Pacific islands. After collecting in the Society Islands in 1857, he developed a close collaboration with Pease. Pease funded some of the field work and Garrett sent him descriptions of the living animals and drawings of shells that illustrated several of Pease's papers. In 1858 the two men considered opening a shell shop in Honolulu, but nothing came of the idea.

In 1860 Pease's life took a tragic turn with the death of Nelly and, a year later, their son. Pease though seems to have recovered remarkably quickly and by the

Fig. 2. Andrew Garrett and one of his south-sea fish. Original photograph in the Museum of Comparative Zoology, Harvard, fish from Garrett 1874

end of the same year he married a Sophie H. True and in 1862 they had a daughter. The next death to be reported was that of his benefactor Haalelea. As a creditor, Pease was given a significant part of the estate and used this to build an extension to his home and to move his collection and library to it. This library contained many books on shells, as might be expected, but also on geography. In addition, he subscribed to several journals and these introduced him to new ideas, such as those on evolution. In September 1860 in a letter to Garrett he wrote: "Darwin's ideas on the origin of species, appears to be absorbing all the attention... In fact so near as I can learn, Darwin has but few backers." Despite the significance of Darwin's work, evolutionary ideas had little impact on the describing of species. In their papers on *Partula* neither Pease nor Garrett discuss the evolution of the species.

Between 1859 and 1863 Garrett's scope expanded. His employers now included the California Academy of Sciences and geographically he ranged widely across the Pacific. He made particularly notable searches of the Kingsmill Islands (now part of the Gilbert islands) and the Society Islands of Tahiti, Moorea, Huahine, Tahaa, Raiatea and Bora Bora. In 1857 and again in 1860-3 he collected in the Society Islands. During these voyages he was distressed at the "indolence, drunkenness, and the most loathsome diseases" that were apparent in the islanders. Despite this impression he was drawn to the islands and was later to settle there.

In Garrett we have for the first time a collector who was meticulous with his data. As he recollected in 1875: "When I explored the islands for *Partula*, I kept a daily journal and gave each species, when discovered, a provisional name, and kept the species from each valley separate. After I sold the collection to Mr. Pease I put up nine sets for him to send to London. He adopted nearly all my provisional names. Carpenter, Adams and Cuming compared the shells with those in the Cumingian collection, and the former published the result of their determinations in the Proc. Zool. Soc, 1864. I reserved a similar set for myself of all except five or six species of which there were no duplicates left after Pease made his selection for his cabinet. Ten years after my first explorations I went over the same localities a second time, so I certainly am pretty well posted on the Partulas of this group."

The specimens he gave to Pease were not so well curated: "Pease was very careless about localities, as any one can see by studying his papers. He was also careless about his duplicates, which he kept in cigar-boxes; and, as he once wrote me, his 'little daughter amused herself in arranging (?) his duplicates.' After his death, those who packed his collection to be sent to Boston must have made sad confusion by the admixture of species." Pease clearly indulged his young daughter and to more than one correspondent commented on her rearrangement of his shells. There is no way of knowing how much she was responsible for the unfortunate confusion of some of the samples, and how much was simply down to Pease's carelessness.

These jumbled specimens should have been sold as an "entire" collection to a Mrs Witthaus of New York for the sum of $3,000 when Pease died in 1871. However, upon inspection she declared that the collection was damaged by careless packaging and refused to buy any but the larger, more attractive shells. The specimens that did not appeal to her were sold to Louis Agassiz, director of the Museum of Comparative Zoology, Harvard University. There they were glued onto pieces of glass or slate for exhibition by the museum's mollusc curator, John Gould Anthony. They were removed from the mounts in the 20th century by William J. Clench. The portion of the collection sold to Mrs Witthaus was eventually donated to the Museum in 1944, reuniting the Pease collection.

Around 1863 the arrangement with Agassiz broke down as the museum was experiencing financial difficulties and suffering the distraction of the Civil War. As a result Garrett shifted his collecting efforts to provide the Godeffroy Museum in Hamburg. In 1870 he moved to Polynesia to live on Huahine, where he married a local woman, Otari, and remained until his death in 1887. During his years on Huahine Garrett published his most significant works, including the highly influential '*Terrestrial Mollusca Inhabiting the Society Islands*'.

His description of his new home (made on his original visit in 1858) gives a view of the landscape of early Huahine: "The Island, in fact, consists of several islets which are separated by narrow channels. They present a bold and mountainous aspect, and are clothed in the most luxuriant verdure from the water's edge to the summits. . . A short distance back there arises an amphitheatre of hills and mountains which are covered either with tall, rank grass or dense dark forests, and the whole

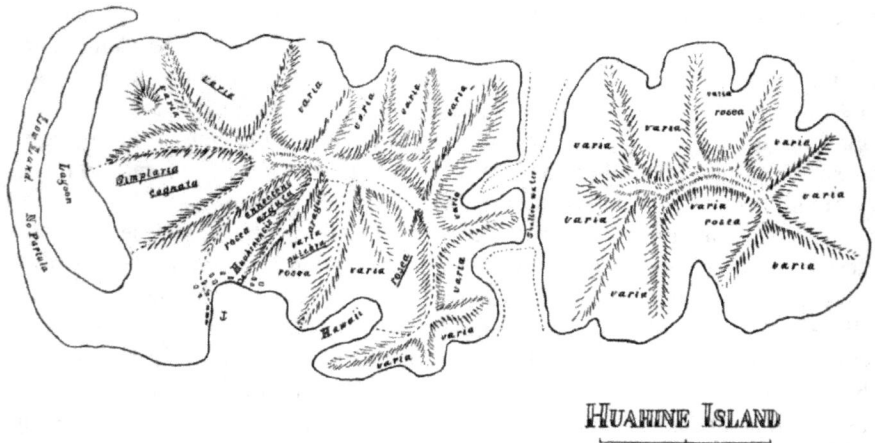

Fig. 3. Garrett's map of Huahine island showing the distribution of his *Partula* species

coast consists of a dense mass of fruit and splendid flowering trees, all combining to form one of the most delightful tropical scenes I ever witnessed." Occasional visits of whalers and traders allowed him access to the main island of Tahiti; the less frequented ones, such as Raiatea and Moorea, were probably reached by small sailing canoe.

There is little information on Garrett the man; when he settled on Huahine he seems to have had a wife, possibly the daughter of the chief of Huahine. His house was a short distance from the village of Fare. It was "a neat little frame house, very comfortable and situated in a nice garden". The location was certainly idyllic, with a view over the island's western lagoon. He was described at this time as a slender man, balding, with side whiskers and a full beard. He looked older than he was; a friend described him as "very unpretentious and no one from casual observation would imagine him to be a savant. . . Outside, his own special study of Conchology, he was deeply read in kindred subjects and no brand of natural history seems to have been overlooked."

In 1887 Garrett died of a mouth cancer. Just before his death he wrote: "For some time I have been unable to work on my collection, which numbers over 8,000 species and 30,000 examples. It is hard to give them up and join the large majority." This collection was acquired by the Bernice Pauahi Bishop Museum of Hawaii in the mid 1890s. Garrett was not exaggerating, and the material that ended up in the Bishop Museum was only a part of what he collected; through Pease, his *Partula* specimens were distributed to the Carnegie Museum, to Philadelphia, the American Museum of Natural History, Harvard and to London. His *Partula* collections must have numbered many thousands of specimens. Modern ethics of conservation and responsibility make this scale of collecting uncomfortable at the very least, but in contemplating it we must remember that the 19th century was a different time. The earth's resources, whether they were minerals, animals, plants, or even people, seemed inexhaustible and conservation of anything was far from the minds of almost everyone.

The vast collections amassed by individuals like Garrett ended up in the world's great museums. Some were broken up by auction on the collector's death but others remained largely intact. Garrett's collection was special as it was made for science rather than primarily curiosity. Although he retained a substantial collection himself, through Pease Garrett's material was distributed to the major museums around the world, where the specimens enabled researchers to identify and classify collections. Identification can be made from the published descriptions for some species, but many of the early descriptions are simply too vague for identification. Although the science of taxonomy had moved on from Gmelin's eleven word description of *Partula faba* in 1791; in the mid 1800s descriptions were often sketchy, for example *Partula arguta* was described in 1864 by Pease as:

"Bulimus shell ovate, narrow, transparent, membranous, thin, narrowly umbilicate, 4 convex whorls, last rounded, suture impressed, mouth wide, ovate; lip simple, reflected; pale straw-coloured. 13 mm long, 8½ mm wide" [Original Latin text: "B. t. ovata, tenuissima, pellucida, membranacea, nitida, anguste umbilicata; anfr. iv., convexis, ultimo ventricoso, suturis impressis apertura ampla, ovata; labro simplici, reflexo ; pallide straminea. Long. 13, diam. 8 ½ mill."]

Although Pease was typical of most 19th century biologists in still describing his species in Latin This practice has since been abandoned in zoology but Latin descriptions remain in botany. In animal descriptions the use of Latin had started to be dropped in the early 1800s, even though a very few individuals were still using it a century later. In 1831 Lesson had dispensed with the Latin for *Partula lutea*. Lesson's description is almost completely useless:

"This species resembles the preceding one [*Partula lineata* – an unhelpful mystery species] but it can be distinguished by a thinner lip, its umbilicus where the opening is obliterated. Its height is 8 lignes [1/12 of a French inch, or 2.26 mm] against a diameter of 5. Its spire is fatter than that of the preceding. The shell is also thinner and entirely a horn yellow colour" (original text in French).

When there were only a few species to distinguish these sorts of descriptions might have been reasonable, but by the time Garrett had finished there were forty species. At this point, fully accurate identification needed the original specimens to be examined directly.

When I collected some of the last Society Island *Partula* in 1992 I needed to identify the species I had found and in order to do so I examined the collection in the Natural History Museum in London. This has the largest zoological collections in the world and has material from Pease and Cuming. The mollusc specimens in the museum are kept in cabinets in an old low ceilinged room. Here thousands of shells are arranged in drawers. The museum's collection of *Partula* is not one of their most extensive, but even so it fills an entire cabinet. There are drawers densely packed with glass vials filled with *Partula* of every colour. There are hundreds of shells with hand-written labels, some with the elaborations of the 19th century, others with more recent functional lettering. Labels by Pease were easy to read, others less so. Labels initialled 'MC' referred to specimens from the Cuming collection. Then there were many with just a locality, often vague, from unknown collectors. Across one species the history of name changes could be traced, but at the same time it was instantly obvious that many had been mislabelled. In many cases over more than a hundred years of adding material from various sources has led to a hopeless confusion. Clearly some collectors had misidentified their specimens and four or

Fig. 4. The Natural History Museum, London, and the collections of the Mollusca Department.

species were labelled as the same, but who had been right? There is only one way to be sure: to examine the actual specimens used when the species was named.

In a separate room in the Mollusca Department are the 'type specimens' upon which the names are based. The Natural History Museum has many *Partula* types, some from Cuming and many sent by Pease. Both men tended to work with large numbers of shells which were sent to museums around the world. Sometimes it is apparent that in packaging up these types they made the occasional mistake, which is confusing. More confusion is added by some specimens being labelled as types when all that was meant was that they were typical, but not necessarily actually part of the type series. In these situations we have to look at as much information as possible: the shells in the different museums, the original descriptions and, in the case of *Partula*, Garrett's careful notes on what he had told Pease before Pease messed up the labelling! Taxonomy is supposed to be a precise science, but encompassing so much scientific history and human foible as it does, at times it can be more of an art. Despite these difficulties and frequent frustrations, one point is clear. Without these collections, preserved for generations, identifying and naming species would be impossible. Without names we would be unable to discuss evolution, and without that the world around us would remain a meaningless jumble of life. The museum collection, and the collector, will therefore always be vital to any endeavour to place life in context.

Today it would be inconceivable for anyone to collect any animal or plant by the thousand as Garrett did with his shells, but although the scale of the material gathered by Garrett is remarkable, what is truly astonishing is that it was to be dwarfed by the collections made in the 20th century.

Chapter 4. The biology of obsession

Partula otaheitana

After the death of Garrett *Partula* research might have been expected to fall into obscurity. In fact there was a brief pause in the collecting, of only 12 years. The next collection seems to have been the 735 specimens collected on Tahiti in a single week by Alfred Goldsborough Mayer in 1899.

Mayer was a marine biologist (specialising in jellyfish) at the Tortugas Biological Station, which he founded as the first marine biological research station in the western hemisphere. He had studied in Harvard under Alexander Agassiz (son of Garrett's patron Louis Agassiz) before becoming an assistant in the Museum of Comparative Zoology. Latterly he was known as Mayor rather than Mayer having changed the spelling of his Germanic name in 1918 to avoid the prevailing anti-German sentiment in America during World War I. At the age of 30 he travelled around the Pacific on the 'Albatross', reaching Tahiti on 6th November 1899. Mayer, along with ensign C.S. Kempff and Dr H.E. Moore, collected in six valleys of the island. In 1902 he published a paper on variation in Tahitian *Partula* based on this collection. Mayer's aim was to investigate how distinct the supposed species were: "A great deal has been written concerning the classification of the species of *Partula* inhabiting the Islands of the Tropical Pacific. Unfortunately, however, the various species have been distinguished only by inspection of the color, form, etc., of the adult shells; and no attempt has been made to dissect the young out from the full-grown snails and thus determine, by direct evidence, whether or no the so-called "species" intergrade, and if so to what extent." In this aim Mayer was taking an approach that had been developed by Ray in the 1600s, evaluating whether the characters used to place an individual in a particular species were inherited from their parents or had developed in the individuals themselves. In this study Mayer dissected out the embryos the snails he collected were carrying, finding that the adults often carried young that differed from their parents in colour patterns or in shell coiling direction.

This was the first truly scientific study of *Partula* and Mayer took a suitably careful approach: "It was my habit to spend the greater part of a day in each of the valleys and to take every snail which was seen." As a result he amassed 735 snails in six valleys in seven days. In ushering in the scientific age to *Partula* evolution Mayer's

Fig. 1. The Tahitian *Partula* studied by Mayer, from his 1902 paper. The bottom rows show some of the embryos he dissected out of the adult snails.

work was to prove highly influential, for it attracted the attention of one of the most obsessive men in the history of science – Henry Edward Crampton.

Partula became an important research group in the early 20th century with the appearance of Henry Crampton. Crampton was an evolutionary biologist, and as such was the second professional biologist to collect *Partula*; like Mayer he had only a secondary interest in the shells.

Crampton's biological experience was very wide, having worked on selection in silkworms and selective death rates in English sparrows and shore crabs. In 1905 he summarised his understanding of evolution in "*On a General Theory of Adaptation and Selection*". In the same year John Thomas Gulick published data he had collected on Hawaiian achatinellid snails back in 1853, concluding that selection was not the cause of the variety of different forms in the islands. Gulick was born on the Hawaiian island of Kauai in 1832 to missionary parents and had collected snails as a child. In his later teens he travelled extensively in the Pacific, returning to Hawaii in 1853. Although Pease was in the islands then, there is no evidence of their interaction even though they must have encountered one another through shell dealing. Gulick does record going to "Orramel's store, looked over my Micronesian shells, and then to Dr. Newcomb's where I remained till after supper. I examined his collection of *Achatinella*, and made some exchanges with him." Newcomb was certainly an acquaintance of Pease as well. Although Gulick collected much of his material himself, some of it was obtained by purchase for on 23rd April 1853 he "bought several hundred shells this morning and have been occupied with them most of the day", again some transaction from Pease seems likely.

It was at around this time that Gulick started speculating that all the forms and varieties of *Achatinella* in his collection represented a continuous series descended from a single species. Although Pease or Garrett could have had the same thought with their *Partula*, they simply did not have the same speculative minds as Gulick. Five years before Charles Darwin and Alfred Russell Wallace first published a summary of their theory of evolution by natural selection, and six years before the publication of the *Origin of Species*, it was clear to Gulick that species were not the immutable creations that most people believed they were. His son, Addison Gulick, a conchologist at the University of Missouri, recalled his father's commenting at the time that "all these Achatinelle never came from Noah's Ark", even though his original aim in the study was to gain insight into God's creative powers. After one year he moved onto other things and did not return to his shells until 1872. In that year he travelled to Britain, hoping to sell his collection to the British Museum. They were not interested in buying it but gave him working space instead. While in London he submitted a paper to the Linnaean Society "*Divergent Evolution through Cumulative Segregation*", claiming that selection could not account for local differentiation. In the same year Gulick had discussions with both Darwin and Wallace. Darwin had favoured selection as the sole mechanism leading to the evolution of new species, but at this time he was coming round to the idea that isolation might also have a role. According to Gulick "the discussion of the subject interested Darwin to such an extent that he invited me to dinner to continue the interview. He referred me to the German author, Moritz Wagner, who had already discussed the subject of isolation. At the close of the interview, he exhorted me not to keep my investigations to myself but to "write, write," he reiterated".

Gulick did publish his conclusions, and his 1905 paper was instantly picked up on by biologists such as Crampton. This paper proved very influential; a quarter of a century later his observations were echoed by the evolutionary biologist Sewell Wright. "The non-adaptive nature of the differences which usually seem to characterize local races, subspecies, and even species of the same genus indicates that this factor of isolation is in fact of first importance in the evolutionary origin of such groups, a point on which field naturalists (e.g., Wagner, Gulick, Jordan, Osborn, and Crampton) have long insisted". This led Wright to propose the concept of random genetic drift and to develop his metaphor of the adaptive landscape.

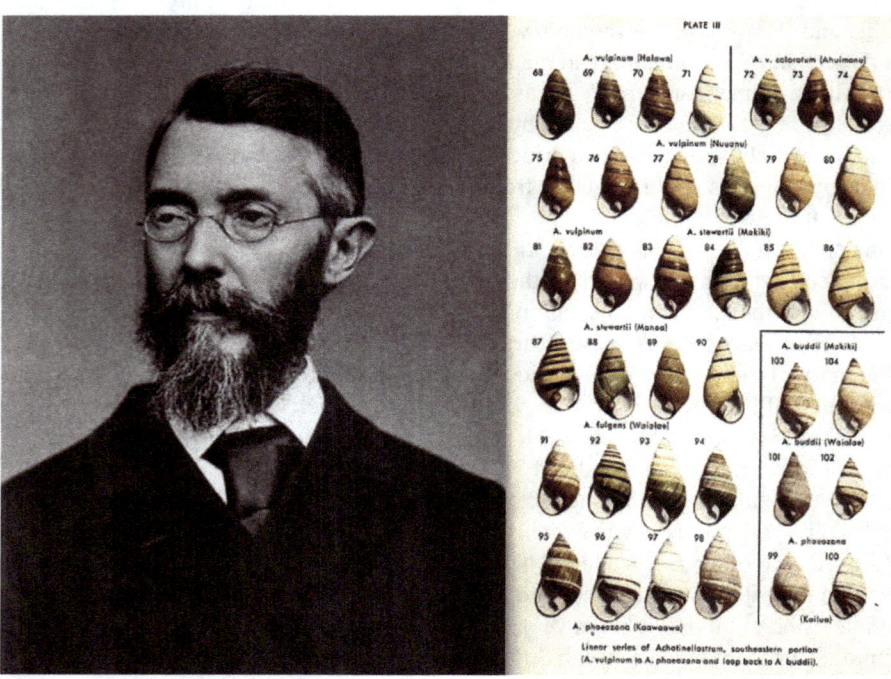

Fig. 2. John Gulick and his *Achatinella* snails. Photo of Gulick in 1889 from Romanes (1897). *Achatinella* from Gulick 1932

Gulick concluded that natural selection on its own, without separation, would not cause divergence and speciation. This was based on the observation that barriers to the movement of species were associated with divergence even in the same habitats. In addition the populations longest separated were seen to be the most different.

Gulick's insights inspired Crampton to start field work in 1906 on *Partula*. Crampton was born in New York in 1875 and received his degree and doctorate at Columbia University, the latter in 1899. From 1893 to 1900 he taught at Columbia, followed by a professorship at Barnard College, which he held until 1943. From 1909 until 1921 he was also curator of invertebrates at the American Museum of Natural History. He carried out field research in South America, West Indies, Polynesia, Micronesia, China, Thailand, Java and Australia.

Crampton belonged to the first generation of biologists who had been educated with evolutionary theory as a core part of their thinking. It is not surprising that he was an enthusiastic 'neo-Darwinist'. These were biologists who recognised

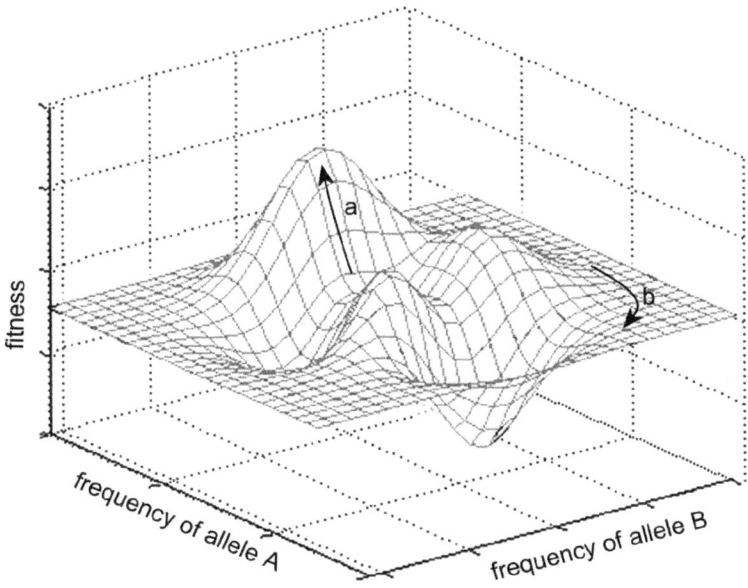

Fig. 3. Wright's adaptive landscape. At point 'a' the arrow shows selection pushing species upwards, towards the 'adaptive peak'. At point 'b' there are no fitness differences between the forms and so no differential selection, here the only changes are random fluctuations causing genetic combinations to change by 'drift'

the opportunities created by what was known as the 'modern synthesis'; the synthesis of Darwin's evolution by natural selection with the elegant heredity of the fledgling science of genetics. One of Darwin's numerous worries had been the mechanism of inheritance. His evolutionary theory relied on organisms passing on their characteristics to their offspring. Some vital essence of the organism must be passed on in sperm and egg to the next generation. So much is obvious, but this would lead to the characteristics being blended with those of the other parent. This was not an insurmountable problem but such blending inheritance would make evolution a terribly slow process. This did not seem right, but Darwin had no choice but to accept it. Darwin published his *Origin of the Species* in 1859, and then sought to back up his theory with research on the inheritance of the characteristics of domestic animals and plants. In 1866 he discussed some puzzling findings with Alfred Russell Wallace. In a letter to Wallace he wrote:

> "I do not think you understand what I mean by the non-blending of certain varieties. It does not refer to fertility. An instance will explain; I crossed the Painted Lady & Purple sweet-peas, which are very differently coloured vars, & got, even out of the same pod, both varieties perfect but none intermediate. Something of this kind, I should think, must occur at first with your butterflies & the 3 forms of Lythrum; tho these cases are in appearance so wonderful, I do not know that they are really more so than every female in the world producing distinct male & female offspring."

Later the same year he wrote to Thomas Laxton:

> "I have spent some hours during the last few days in examining with the greatest possible interest your Peas. After all, I could not resist your very kind offer, & have kept one pea out of the packets marked 7. 10. 11. 12. 13. & the purple pea No. 14. I never saw anything more curious than the lots 12 & 13. Will you have the kindness to look at the pod of No. 13, & you will see that the rim close to the suture is red; pray tell me whether you think this has been caused by the pollen of the purple pod. — These two cases & that of the purple pea are truly wonderful; the others are less striking. But I observe in lot 1 (except one pea) & in a lesser degree in 2, & in 3. 5. 6. & 8 that the crossed peas are smooth like the paternal stock, & not wrinkled & cubical like the mother pea— Can this loss of wrinkling be due to mere variation, or to the effect of some peculiar culture, or is it the direct result of the pollen of the father? I should be grateful for an answer on this head— I know I am rather unreasonable, but I should be very much obliged if you would write a single word in answer to 3 queries on the enclosed paper."

Fig. 4. Henry Crampton photographed in 1927 or 1934, from the *Embryo Project Encyclopedia*. ISSN 1940-5030 http://embryo.asu.edu/handle/10766/2917

Darwin was working on too many fronts to give this observation of a lack of blending inheritance in his sweet-peas the attention they deserved. In Austria the monk Gregor Mendel had no such distractions and in this same year had just published his work on exactly this topic. He had been breeding sweet peas since 1856 and by 1863 had the amassed data on the characteristics of 12,835 plants. These he reported on to the Natural History Society of Brno in 1865, which published the reports the following year. Mendel's results were much the same as Darwin's but on a much larger scale and precisely quantified. He was able to use his data to show that the pea plants passed on what he called "factors"; discrete hereditary units that we now know as genes. In the plant they occurred in pairs, but only one was passed on by each parent, so the offspring contained a mixture of the genes of the parents, but not a blend. In his plants one form ('allele') of the gene was dominant and one recessive, the dominant characteristic masking the recessive.

At the time of its publication no-one realised its importance. It was thought to be just about hybridisation and even Mendel thought it only related to some species or categories. In correspondence with Mendel the Swiss botanist Karl Nägeli told him his work was too preliminary to be very meaningful and persuaded him to extend its scope by taking the same approach with hawkweed. Frustratingly, the results were inconclusive for, unknown to Mendel and Nägeli, hawkweed reproduces asexually as well as sexually, confusing the inheritance pattern. This appeared to disprove the universality of Mendel's inheritance and when Nägeli published his great work, *"A mechanico-physiological theory of organic evolution"* (1884), he ignored Mendel completely. In this book he speculated that inherited characters were passed on through part of the cell's contents that he called the 'idioplasm'.

One of Nägeli's students was Carl Correns. Under Nägeli's influence Correns also turned to hawkweed as a model plant. Naturally therefore, he was familiar with Mendel and had read his work on peas. In 1900 Correns read a paper by Hugo de Vries on inheritance and found some startlingly familiar results in it. De Vries had been working on heredity for some time and in 1889 had published *"Intracellular Pangenesis"*, modifying Darwin's view of inheritance to suggest that different characters were passed on by different carriers of heredity and that these were particles of some form; these he called pangenes, later to be shortened to genes. In order to investigate this he carried out breeding experiments. His results were similar to Mendel's and with his interest in statistics de Vries noticed that a simple explanation could be proposed based on binomial probability, with the characters being controlled by two different 'pangenes'. These conclusions he published in his 1900 paper. Correns spotted the similarity with Mendel's conclusions, as a result both De Vries and Correns are credited with the rediscovery of Mendelian genetics.

De Vries went on to make one more vitally important contribution to evolutionary biology, that is the recognition of mutation as a source of variation. All

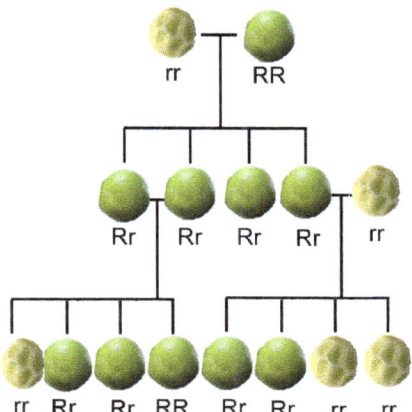

Fig. 5. Mendelian inheritance of wrinkled peas, the smooth form is dominant (with allele combinations of RR or Rr) while the wrinkled form is recessive (rr)

prior studies had been concerned with the maintenance and inheritance of variation, but had not considered its origin. 'Mutation' itself was a term invented by De Vries to describe the sudden appearance of completely new varieties. For his part, Correns turned his attention to the situations where Mendel's laws did not seem to fit. Through experimentation he concluded that in some species the sex of the parent was also significant, that is there is a 'maternal effect' whereby the non-genetic contents of the egg, that is those provided exclusively by the mother, determine a characteristic of the offspring.

With the publicity given to Mendel by Correns in particular, Mendelian genetics rapidly spread out of the confines of hybridisation experiments in botany. Within four years the Swiss zoologist Arnold Lang had published an explanation of the genes controlling some of the colour forms in the banded snails *Cepaea hortensis* and *C. nemoralis*. The demonstration that genetic experimentation was possible caught the attention of Crampton.

Crampton's research had originally been on embryology, with his first publication (1894) being on the reversal of cell division patterns in a sinistral snail (a species with a shell coiling to the left, or in an anti-clockwise direction, instead of the more usual right, clockwise, or dextral direction). He did not initially see this as being of major importance, although he was to return to the topic of coiling, or 'chirality', two decades later. He carried on working on embryology of snails and sea squirts until 1900, also branching out into development and variation in saturnid moths. He started writing on evolution, publishing the book "*The Doctrine of Evolution*" in 1924. His moth work was particularly concerned with variation and he wished to compare his findings in moths with another group of animals. Alfred Mayer's 1902 paper on Tahitian *Partula* caught his attention. It was Mayer's conclusions that inspied Crampton: "These color-sports tend to breed true to themselves, and therefore to originate new color-forms and finally new species. This tendency is, however, held in check by frequent inter-crossing with the parent stock, and becomes effective only when the new color-variety is isolated, or when it displays a remarkably strong tendency to breed true." Here was an indication that Mendelian genetics could be applied to tree snail banding as Lang had done with *Cepaea*. This inspired Crampton to take on Pacific tree snails. He planned to work on the fantastically varied Hawaiian Achatinellidae but a discussion with the Hawaiian conchologist Dr. C. Montague Cooke Jr. and with Mayer led to him change focus to *Partula*.

Crampton started investigating evolutionary processes in *Partula* in 1906. He secured funding from the Carnegie Institution and started his collecting on Tahiti and Moorea in 1907, collecting 6,445 snails on Moorea, before moving on in 1908 to the Leeward Islands of Huahine, Raiatea, Tahaa and Borabora.

Identifying *Partula* required an examination of the described species in the museums, and determining what was already known of their distribution required examination of Garrett's collection in particular. This was now in the Bishop Museum in Hawaii and in 1909 Crampton visited Hawaii. There he met Cooke. Cooke, now 35, had recently been appointed Curator of Pulmonata in the Bishop Museum. Cooke's research interests were the Hawaiian achatinellid tree snails and the tornatellinids. One notable feature of Cooke's own collecting was that he took great care to preserve the bodies of the snails as well as their shells. This was a practice that Crampton adopted, although his own work was almost entirely concerned with the shells.

Examining Garrett's collection Crampton realised that some sites needed more collections, so he returned to Polynesia in 1909 to collect at a different time of the year, collecting 3,304 on Moorea and Raiatea. By this time he had collected over 80,000 snails from over 200 valleys in the Society Islands.

The 1916 Tahiti volume, the first of his three monographs, covered the seven species already recorded from the island and described several varieties (he was to describe one more Tahitian species 14 years later). His primary aim was to elucidate the genetics of these species and he paid close attention to the similarities between the embryonic shells and their parents. *Partula otaheitana* proved to be a particularly useful species, as Mayer had found earlier. Occurring throughout the island in a variety of colour forms and patterns and with left- and right-handed spirals, there were many characteristics that could be investigated. In the most widespread subspecies, *P. otaheitana rubescens*, Crampton reckoned all snails could be divided into two main colour forms: red (including orange) and yellow. After excluding valleys

Fig. 6. One of the plates from Henry Crampton's Tahiti monograph

where only a few snails were found or where only one colour form occurred he determined that of his adults containing embryos, 46% were red but only 37% of their embryos were this colour. He proposed that the inheritance of red and yellow colours could be described in terms of two genes which he termed D and R, for dominant and recessive, with red being the dominant colour and yellow the recessive. In this he was using a modified labelling system which has since been dropped and we have now reverted to Mendel's original system of labelling dominants with an upper case letter, and recessives with a lower case version – thus Crampton's colour alleles would more usually be labelled R and r. Accordingly, I have modified Crampton's discussion by using the current labelling convention. As the terminology implies, the dominant is the stronger gene and the recessive can only be expressed when it is in isolation. A snail inheriting a recessive gene from each of its parents (a so-called 'homozygous recessive') would have two such genes and be represented as rr. This would be a yellow snail. A 'homozygous dominant' snail would be RR and, predictably, red. A snail inheriting a dominant gene from one parent and a recessive from another would be Rr, a 'heterozygote'. As red was dominant the red (R) gene would be expressed and the snail would have a red shell. This interpretation worked reasonably well in some valleys, but in others there seemed to be a mismatch, with 22% more of one colour than he expected. He suggested that this difference might be due to the red colour developing later in some individuals, so there were no real differences between adults and offspring.

The banding pattern also seemed to fit a Mendelian pattern with banding being produced by a recessive gene. He repeated his analysis with each subspecies and this seemed to show that colour and pattern variation in all *otaheitana* subspecies was controlled by the same genes. Given that they were all very similar and shared

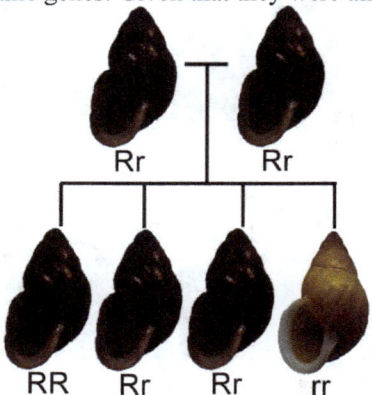

Fig. 7. Mendelian inheritance of colour in *Partula otaheitana* where the red allele (R) is dominant to the yellow allele (r)

the same origins, this can hardly have been a surprise.

After completing the Tahiti monograph he moved onto his Moorean samples. Realising that some areas needed more attention, he collected on the island again in 1919. Although these collecting trips were often arduous in the extreme terrain of the islands, he found them enormously rewarding:

"Almost without exception... the exploration of a valley can be accomplished only on foot, owing to the steep declivities to be traversed, the deep streams to be forded, and the absence of any trails whatsoever in the thick forest and undergrowth of the areas inhabited by Partulae.

"It was my custom to start in the early morning with two or three native assistants, more rarely alone, and to spend the greater part of the day in the interior. Full notes were made as to the character of the vegetation, the favored nurse-plants, the topographical features, the inland distances and barometrical levels of the points where snails were first met and their headquarter in a given valley, as well as other pertinent facts."

"Suffice it to say that the days and nights of arduous and sometimes dangerous effort included hours of keen enjoyment, for the island of Tahiti, especially, is of matchless beauty, while the chiefs and their families offered abundant hospitalities which it was a privilege to enjoy at the time as it is now a pleasure to acknowledge them."

His focus on Moorea was interrupted briefly by a survey of the Marianas in 1923. The small fauna and simple topography of these islands meant that it took him just two years to produce the Marianas volume on his 5,575 snails from Guam and 3,360 from Saiapan. In contrast, he made extremely slow progress on the Moorean monograph. Each analysis he started seemed to demand more data and he returned to the island in 1923, and again in 1924, still not content with the level of detail of his samples.

It took Crampton 16 years to reach a point at which he was satisfied with his Moorean work. By the time of its publication in 1932 the volume covered 116,166 shells of 10 species. The slow progress in publishing this monograph was the result of a more complicated situation as well as Crampton's increasingly obsessive need for more and more measurements. Although there were not very many species on the island there was more variation in patterns than on Tahiti, and some of the species could commonly be found as right-handed (dextral) or left-handed (sinistral) shells. On Tahiti most species were one or the other, with only rare reversed coiling individuals. On Moorea different valleys usually had shells that coiled in one particular direction but these could vary between adjacent valleys and in some valleys they

Fig. 8. The mountainous terrain of Tahiti. Crampton's collecting was carried out predominantly in the valleys. These supported the largest populations of *Partula* and he thought they did not occur above 600 m above sea level. In fact, although ridge populations were more sparse, they did exist. (photo: P. Pearce-Kelly).

they were mixed. In fact, most valleys showed some mixture, even if they were dominated by one form. Naturally, as he knew a great deal about this issue from his early experimental research days, this attracted Crampton's attention and he applied his now well-tried technique of comparing adults to their embryos. However, the results were not what he anticipated:

> "This last case is excessively important in connection with the supposed determination of the direction of coil by "maternal inheritance" a topic first brought up by the single instance of a *mirabilis* which contained young of the two kinds at the same time. Here it is an unusual dextral mutant of a sinistral species which presents the noteworthy facts, while the *mirabilis* was a sinistral mutant of a typically dextral species. Other cases will be described in *P. suturalis*. Collectively they exclude the hypothesis of maternal inheritance as a general one, according to which all young would be somatically [in appearance, rather than in genes] like the bearing

parent, whatever their own chromosomal factors might be."

It is clear from this that Crampton had become hopelessly confused. Whereas he explains the development of some patterns logically, even if somewhat convolutedly, shell-coiling on Moorea led him round in circles. The simple explanation that he might have been hoping for was the dominant-recessive pattern that worked so well for colour and banding pattern. This simply did not work: his data would not fit a neat Mendelian dominant-recessive pattern, which left the alternative, the new hypothesis of 'maternal inheritance'. In 1923 Boycott and Diver had published three years of work on some sinistral variants of the pond snail *Lymnaea peregra*. They bred over 16,000 snails but could make little sense of their results – same coiling pairs seemed to produce almost all combinations of offspring. On reading their paper Alfred Sturtevant noted the similarity between the *Lymnaea* results and those on Tahitian *Partula* by Mayer and Crampton. He realised that the strange results were explicable as "An analysis of the data presented suggests that the case is a simple Mendelian one, with the dextral character dominant, but with the nature of a given individual determined, not by its own constitution but by that of the unreduced egg from which it arose." Sturtevant was basing this on the *Partula* observations and, as such, was making a guess. It was however, an inspired guess as he was

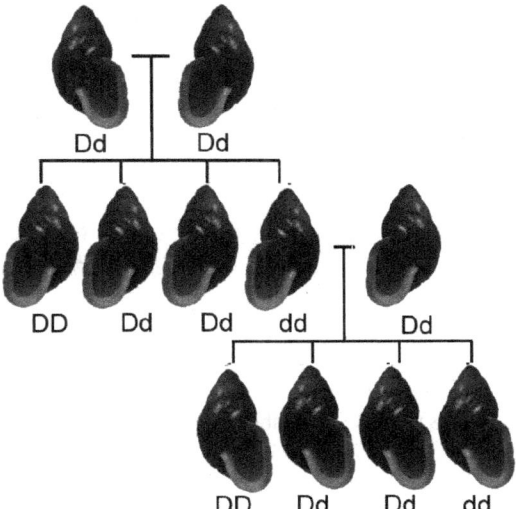

Fig. 9. The maternal effect in *Partula otaheitana*, where sinistral coiling (D) is dominant. The individual acting as female is shown on the left of the pairing. The dextral genes (d) can only be expressed in the homozygous recessive individual (dd) which then produces proteins in its eggs, leading to dextral offspring.

proved right by further research by Boycott and Diver, among others, but although they published their findings in 1930 Crampton did not manage to incorporate the understanding of chirality into his Moorean monograph. What Sturtevant had hypothesised and what the 1930 research found, was that dextral and sinistral forms were inherited in a predictable Mendelian fashion but with a complicating factor, the 'maternal effect'.

This complication was that the spiralling of the embryonic shell was determined by a factor in the egg. This factor was produced by the parent, and not the embryo itself, hence the 'maternal effect'. This means that spiralling is controlled by genes, but the mother's genes, not the embryo's. Even though it has a Mendelian pattern of inheritance, it is delayed by a generation. The 'maternal effect' was a new idea at the time and it would not be thought through to its conclusion for many years. Hence, explaining spiralling (or 'chirality') in *Partula* would not be achieved until the next stage in the story.

Crampton's interpretation of the evolution of the Tahitian and Moorean species was that they could mostly be grouped into species pairs with one species on each island: *otaheitana* and *mooreana* (Tahiti and Moorea respectively); *nodosa* and *suturalis*; *filosa* and *taeniata*; *attenuata* and *exigua*. One Moorean species, *tohiveana*, seemed to be related to the Raiatean species *formosa* and *dentifera*. This suggested to him that there might have been an ancient land connection between Moorea and Raiatea. The presence of *Samoana attentuata* on more than one island indicated to him that the islands had subsided rather than being separate volcanic extrusions. Geology has since refuted this idea and the Society Islands are now known to represent a chain of islands moving westwards from a volcanic hotspot. This is obvious when looking at the islands: the youngest island is Tahiti (800,000 years for the oldest rocks, some parts much younger), and this is clearly an extinct volcano. Its central crater has broken so that a deep valley cuts through the middle of the island, but its basic shape is still apparent. Nearby Moorea is around a million years old and, although its crater rim has fragmented more, is also clearly a remnant of a volcano. The next youngest island, Huahine (2.3 million years old) is more eroded and the slightly older Raiatea and Tahaa (2.5 million years) are lower, more eroded and have lost the classic volcano shape. Further to the west 3 million year old Bora-Bora is just a spike of lava set in the centre of a coral atoll. It is the last stage of an old volcano before wind and rain grind it down completely. 4 million year old Maupiti, further westwards, is in the same stage. Further westwards lie the next stages where the volcano has disappeared completely, leaving a ring of coral rising above the sea, these atolls of Motu One and Mamuae are slowly submerging as their underlying volcanoes collapse. The geological plate that the Society Islands rest on has been moving westwards at a rate of around 10 cm a year, moving over the volcanic hotspot which now lies around

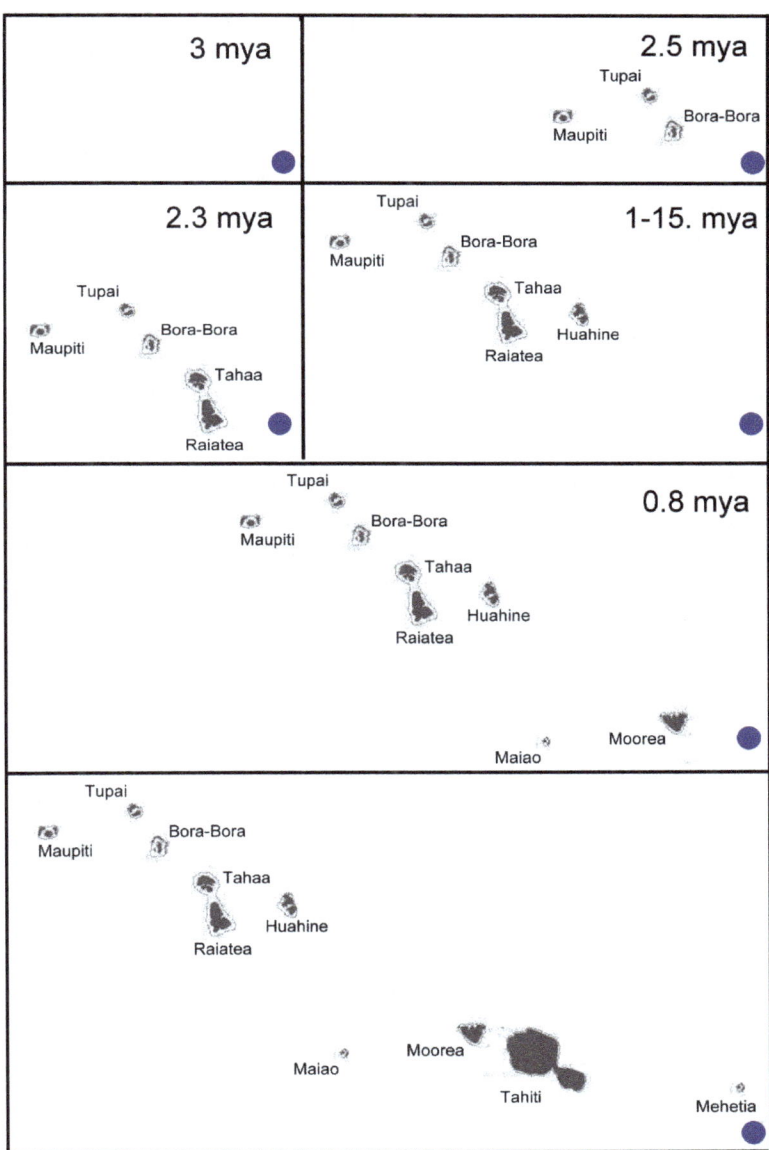

Fig. 10. Geological evolution of the Polynesian islands, showing the fixed position of the volcanic hotspot (the shaded circle) and the islands drifting northwestward. Approximate dates are given as millions of years ago (mya).

100 km to the south-east of Tahiti. There it can be seen in the undersea volcanoes of Tehitia and Moua Pihaa and the island of Mehetia. These are the latest in the sequence of volcanoes being thrust up over the hotspot and then dying as they drift off to the west, slowly eroding into the sea.

This progression of older islands moving westwards means that there are two opposing processes present. On one hand the more time an island has been occupied by snails, the more the species may have diversified on it, meaning that older islands have more species than younger ones. On the other hand, the older the island, the more eroded it is. As it shrinks under the actions of wind and rain it becomes less able to support many species. So as an island ages, the more its species diversify, but also the less able it is to support them. If we look at *Partula* there is a general pattern of increasing numbers with island age: from nine species on Tahiti, and eight on similarly aged Moorea but 34 named species on much older Raiatea. Huahine, lying between the two has only three species but is tiny. Even if the sizes of the islands are taken into account Raiatea seems to have far more species than it should do. Bora Bora, oldest and smallest of the *Partula* islands has just a single species. Why though does Maupiti lack *Partula*? It still has a mountain and is the oldest of the lot. Presumably it is just too small, at a mere 11 km^2 it is less than half the size of Bora Bora, in fact the volcanic island is even smaller than that, a good half of the area being the off-shore coral atoll.

Another aspect that Crampton spent much time considering was the changes that might have taken place in the half-century since Garrett had collected on Moorea.

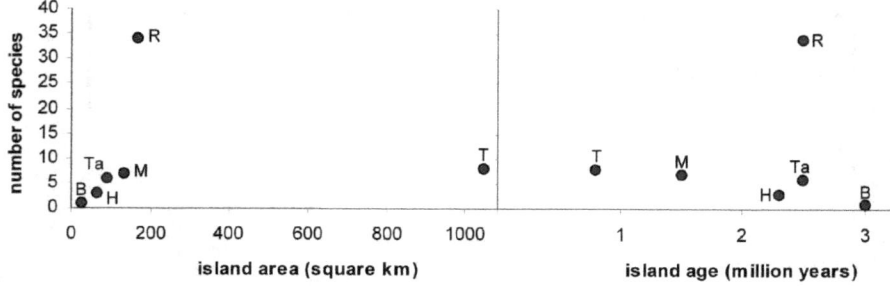

Fig. 11. The number of *Partula* species on different sized islands. Note that Raiatea has far more species than expected from either its size or age, suggesting that many of the 'species' on that island are probably subspecies or local varieties. B – Bora Bora; H – Huahine; M – Moorea; R – Raiatea; Ta – Tahaa; T - Tahiti

"When my own material is broadly compared with the collections made by Garrett from 1860 onward, marked differences are seen in many ways, showing that changes had taken place in some species at least since Garrett made his observations and prior to 1907... The case of *Partula mooreana* is very clear, for it is certain that this species has only recently migrated to the present limits of its range in Tefeo and Roroie Valleys, across the divide, and to Haapiti Valley to the northwest, from the original headquarters in Vaianai Valley. The variety *Partula taeniata elongata* has also reached new-won territory in very recent years. The whole history of *Partula suturalis vexillum* is by far the most important because in this case the evidence is most conclusive that the variety has reached the valleys of the extreme northwest and northeast since Garrett's time, while in addition the intrinsic evidence of the material itself proves that a prior general condition of sinistrality is steadily giving place to one of prevailing dextral nature. The histories of *Partula mirabilis* and *Partula aurantia* are also important in the present connection, mainly, however, as supplementary instances of recent enlargement of the territories of occupation."

This issue of *Partula* species expanding their ranges was to be revisited 40 years later, with very different conclusions.

By this time the Carnegie Institution was becoming frustrated by the never-ending field work and was pressing Crampton to move to laboratory studies. Crampton's inspiration, Alfred Mayer, also complained: "I wish he or someone would ... *stay* in the Islands long enough to do *breeding* experiments. Crampton only nibbles around on the *outside* of the problem and can never solve it by his present methods of mere collecting." The Carnegie Institution offered to fund annual visits to monitor experimental transplanted colonies but not for further collecting. His unwillingness to accede to their wishes led to a parting of the ways. Through the auspices of Cooke, from 1929 the Bishop Museum replaced the Carnegie Institution as his financial backers. The Bishop Museum was becoming increasingly involved with *Partula* though Cooke who had dedicated much of his professional life to making the museum's collection as comprehensive as possible, adding nearly three million specimens. This was through exchange of specimens, collaboration with researchers like Crampton and through new collecting expeditions. One of the most important of the expeditions he led was the Mangareva Expedition of 1934.

The Mangareva Expedition called into Polynesia on its way back to Hawaii. There they surveyed Tahiti, Moorea, Huahine, Raiatea, Tahiti and Bora Bora. The highlight of the collecting was finding the very rare *Partula turgida* on

Raiatea, this had been one of their sought after species and Cooke offered a 15 Franc reward, raised to 25 on the last day. At the last minute they found a small colony of 14 adults and 47 juveniles.

Among the crew of the fishing boat 'Myojin Maru' chartered for the Mangareva Expedition was chief-engineer Yoshio Kondo, a 24 year-old Hawaiian. Within weeks of the start of the expedition Kondo was also acting as a regular collector and on the return to Hawaii he was appointed Assistant in Malacology, and at the same time studied biology at the University of Hawaii.

After completing the Moorean volume Crampton had been working on the Raiatean species, collecting again in 1935. Snail research suddenly became a low priority with the outbreak of World War II and work in the Pacific was an impossibility for several years. Crampton could not keep away for long and was back on Raiatea in 1947. In the same year he was visited in New York by Cooke and over the course of a discussion of Partulidae Cooke raised the issue of his impression that the high-altitude species on different islands were remarkably similar. He wondered whether there might be a deeper similarity and relationship between these species which might be reflected in anatomical similarities. Crampton's experience was almost entirely based on shell characters and he doubted that anatomy could tell them anything worthwhile; as Kondo described, he expressed "forthright skepticism [sic]" over the idea. Despite this, Crampton and Cooke together came up with the idea of an anatomical study. This was to be part of a grand study, the comprehensive survey of the anatomy of the endemic Pacific snails that was being undertaken by Cooke, Clench and Kondo. Cooke was one of the leading proponents of the need for anatomical studies to refine shell-based classifications. Cooke had been curator of the snail collection of the Bernice P. Bishop Museum since 1902 and had bought Gulick's important collection for the museum in 1905. Between 1911 and 1920 he had been involved with the Achatinellidae section of the greatest mollusc taxonomic work of the early 20th century: the *"Manual of Conchology"* which had been published in multiple volumes from 1879. This work was led by the most prominent mollusc experts of the time: first George Washington Tyron and then Henry August Pilsbry. Pilsbry in particular was concerned with revising the classification of all molluscs based on their anatomy, and not just their shells. This made the *"Manual of Conchology"* a revolutionary work. It was Pilsbry himself who for the first time used anatomical differences to split the Partulidae into three genera: *Partula, Samoana* and *Eua*, in the 1909 volume of the *Manual*. Working with Pilsbry was a major influence on Cooke's approach to classification.

Under Cooke's supervision, Kondo started work on the Partulidae in 1947 but Cooke did not see the results, dying the following year. Cooke's death interrupted the project as Kondo took over his post at the Bishop Museum and administrative duties intervened. By 1954 he was sufficiently settled to take it up again, now working on it for a PhD at Harvard University, under the supervision of William J. Clench. By 1948 he had already dissected 60 species and the work was completed in 1955, after the dissection of a total of 107 species.

Kondo found that the Partulidae could be divided into a number of species groups based on their reproductive anatomy, and in so doing convinced Crampton that anatomical study was not a waste of time after all. This grouping based on reproductive anatomy was to be expected as it was known from other snails and it was probable that closely related species that coexisted would have some mechanism to prevent hybridisation. 'Allopatric' species, that is those living in different places, such as *Partula* on different islands, would not need such mechanisms and so may remain very similar. 'Sympatric' species, living in exactly the same places, would hybridise unless something made hybridisation unlikely. In birds plumage colouring and song serve this function, but in deaf and almost blind snails a more fundamental barrier must exist. For this reason most sympatric snails, even if identical to human eyes, can still be distinguished by their reproductive anatomy, principally by penis shape. This affects the likelihood of successful mating, and these differences are exactly what Kondo found. However, it was not quite as simple as was expected; in general a species showed a particular anatomical form but in a small number of cases one species might have two or even more anatomical forms. He did not know what to make of this, but it certainly seemed an interesting point. By this stage Crampton was coming round to the view that anatomy could make a useful contribution to understanding partulid evolution. He himself was making ever slower progress in completing his monographic series and finally time ran out. He died on February 27th 1956, at the age of 81.

Fig. 12. Yoshio Kondo. Photograph by T. Ross (1967)

Kondo continued to work on partulids, but only sporadically. His anatomical work was never published and he would probably not have made any further contributions had he not started collaborating with John Bayard Burch in the 1960s. Burch, Curator of Mollusks at the University of Michigan Museum of Zoology, has interests in all aspects of mollusc biology and in 1970 he collected extensively on Tahiti, Moorea, Raiatea and Bora Bora along with his son John B. Burch, Jr., Evin Oshima and C.M. Patterson. This was part of a project in collaboration with Kondo to conduct a comprehensive anatomical study of all partulid populations of Tahiti. Some studies did emerge from their collaboration, in particular a return to Cooke's old ideas about the high-altitude species. One of Burch's specimens was particularly distinctive; a high altitude *Samoana* species from Tahiti which Kondo described as *Samoana burchi* in 1973. In the paper describing the species Kondo concluded the thought process that Cooke had started with Crampton in 1947: in 1955 Kondo had reclassified *Partula attenuata* as a *Samoana* based on its anatomy and now he found that all the thin-shelled, high-altitude species belonged in *Samoana*.

Several years later Kondo picked out some more of Buch's specimens. Some specimens from a population of *Partula otaheitana* were anatomically very distinctive, and looked more like *Samoana* than *Partula*. In 1980 Kondo named it *S. jackieburchi*, after Burch's son. For a *Samoana* it was odd in having a thick, *Partula*-like shell, but Kondo was confident of its identity based on his dissections. Opinion was to change on this issue and genetic data led to its transfer to *Partula* in 1986. Quite what *jackieburchi* is remains an enigma: could it be a thick-shelled *Samoana*, is it a *Samoana*-like *Partula*, or just a group of abnormal *otaheitana*?

Kondo and Burch's collaborative project was never completed. Kondo retired in 1980 and did not publish anything further on partulids before his death in 1990. He had worked up what had been most interesting to him and the project with Burch was never completed. The snails that they had collected were sent alive to Michigan in the 1970s. There they were freeze dried and stored in the museum before being forgotten about for 40 years.

Chapter 5. The Three Professors

Partula suturalis

The modern genetic age reached *Partula* in 1962 with the arrival of two young geneticists: Bryan Clarke and Jim Murray, and later Michael Johnson. All three later became notable Professors of genetics, at the Universities of Nottingham, Virginia and Western Australia.

Bryan Clarke and Jim Murray were carrying out doctoral research on variation of the banded snails *Cepaea nemoralis* and *C. hortensis* in Oxford in the late 1950s and early 1960s. Their DPhil theses were on: "Some factors affecting shell colour polymorphism in *Cepaea*" (Clarke 1959) and "Movement and mortality in a colony of *Cepaea nemoralis*" (Murray 1962). This research followed on from the work of Arthur Cain and Phillip Sheppard and investigated

Fig. 1. The Three Professors as they became known: Bryan Clarke at the site of habitat experiment on Moorea 1982, Jim Murray and motor scooter on Huahine 1987 and Mike Johnson on Raiatea 1980. Photos: J. Murray.

The selection pressures that produced colour variation in the snails. Cain and Sheppard were part of the "Oxford School of Ecological Genetics" founded by E.B. Ford. Cain and Sheppard took on banded snails because polymorphism had been of evolutionary interest for a long time and these were conspicuously polymorphic, but mainly, according to Cain, because when "I poured out on the table in front of him [Sheppard] a sample of *Cepaea nemoralis* shells... and we decided then and there (a) that it was impossible that such striking variation could be wholly neutral, and (b) that we would work on it". They concluded that there was a "physiological advantage of certain gene combinations" producing a "balance of different genetic forms in each population", this balance being further altered by the actions of predators.

In 1962 Clarke reanalysed Crampton's data from 1932, paying close attention to the valleys where several species showing the same polymorphisms coexisted in mixed colonies in Moorea. In the species *Partula suturalis* and *taeniata* the percentage of banded forms is negatively correlated; where one species was often banded the other was rarely so, showing that variation is not random,

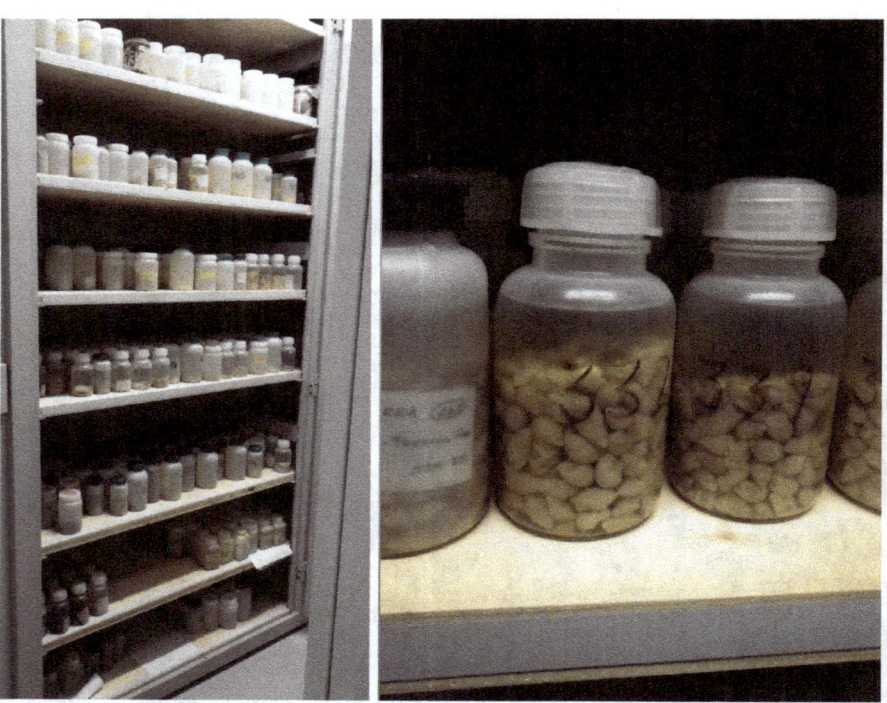

Fig. 2. The *Partula* collections made by Bryan Clarke and Jim Murray, stored in the Natural History Museum, London

contrary to Crampton's views, and suggesting that selection was in play. In this paper Clarke explained the maintenance of polymorphisms as a result of predators forming a search image which helped them recognise the common forms and so giving an advantage to the rarer forms: "I shall call them *apostatic polymorphisms* because the crucial thing about them is the selective advantage of phenotypes that stand out from the norm; in other words, the selective advantage of apostates." This 'apostatic selection' maintains a great deal of variation in different species, most obviously in many different sorts of banded *Cepaea* snails. These may be yellow or brown, with two, four or more bands, or unbanded. In woodland areas brown forms survive best and are most abundant, and in grassland yellow forms usually predominate. However, no one colour or banding form excludes the others. This is because of apostatic selection: thrushes feed upon the snails, finding the commonest form and developing a 'search image' which enables them to recognise the snails and detect them quickly. Over time they eat most of the snails for which they have a search image, leading to them becoming increasingly rare. Eventually their preferred form is so rare that they are unable to find them and encounter other forms, developing a new search image. The rarest forms, those that the thrushes are not looking for, are at a selective advantage – hence, Bryan Clarke's 'apostatic selection' and so the snail population is constantly moving between different colour forms, and remains polymorphic. Clarke also noted that many of the obviously polymorphic snail species are tree climbers, perhaps being more conspicuous to predators.

Partula had received some attention after Crampton's work when in 1956 Donald Bailey re-analysed the data to investigate the evolutionary distinctiveness of different geographical varieties of *taeniata*. Bailey concluded that they were diverging as a result of reproductive isolation between subspecies. Quite what caused that isolation was unknown and Clarke concluded that "We cannot speak conclusively until more is know about the ecology of *Partula*, but a likely explanation for the diversification of subspecies is that there has been selection acting directly upon the characters studied by Bailey, and that this has operated without necessary differences in ecology. In other words: apostatic selection has been at work."

Clarke and Murray started their own field research on Moorea in 1962. Crampton's conclusion that *Partula* variation was non-adaptive was essentially a direct contradiction of the *Cepaea* work. The beautifully described variations in Crampton's monographs provided an excellent resource for the investigation of the patterns in this group. A few years later Michael Johnson joined Bryan Clarke's laboratory as a postdoctoral researcher, working on the familiar *Cepaea* banded snails, making use of the new biochemical techniques to identify genes. Whereas Clarke and Murray had always worked on snails and had spent many years inferring the presence of genes from their effects on the animals, Johnson's earlier research had been on the biochemistry of tobacco mosaic-virus and on enzyme variation in fish. He was soon drawn into the *Partula* research and for the best part of 20 years Clarke, Murray and Johnson were to be a highly productive team, despite securing Professorships in as widely spaced locations. In their years of working on *Partula* they investigated speciation and the evolution of specific characters, such as colouration and chirality.

Investigating selection on colour involved field recording of the distribution and abundance of different colour forms, just as Crampton had been doing with *Partula*. They went further though, with laboratory breeding of snails enabling them to cross colour forms and work out the interactions of the genes that controlled the patterns. They developed a simple but effective way of keeping the snails: using plastic boxes lined with damp paper, kept at constant temperatures. The snails were fed an artificial diet of oatmeal, lettuce and powdered chalk. Their techniques for breeding *Partula* were unexpectedly to prove vitally important in the future.

In the 1960s though, the breeding had one specific aim – to enable different gene combinations to be crossed. Different colour or spiralling forms were kept in pairs and the characteristics of their offspring recorded. In this way they were able to determine how the patterns could be inherited. For example, when crossing *taeniata* snails with white lips to their shells with those with coloured lips and then crossing the offspring with one of the parental forms (backcrossing) it was found that the offspring of a cross between pink and white were all pink. Crosses between these offspring (the so-called second filial generation – F_2) sometimes recreated the white lip. This is explicable by lip colour being controlled by one gene with two forms (alleles): a dominant pink L^p and a recessive white L^w. Thus pink lipped shells have at least one L^p gene, so are either homozygous dominant (L^pL^p) or heterozygous (L^pL^w), where the pink gene dominates the white. White-lipped shells can only be produced by snails completely lacking pink genes (homozygous recessives: L^wL^w). This is exactly the same pattern as Mendel had recorded with his peas in the 19th century. Other patterns seemed to be more complicated, with more than one gene contributing to the colour. For example, general shell colour was controlled by two different genes, one relating to yellow-brown colour and one to pink shades. For the yellow-brown gene the brown allele was dominant, but the yellow allele was modified by the pink allele, giving a range of shades from dark brown through to a pink tinted yellow.

Extremely complicated interactions were found in the banding pattern of *taeniata*. Here Crampton form 'lyra', a shell with three dark bands and a dark area above the upper band, could also produce the double banded brown 'zonata' form and unbanded shells. This could be produced by one gene with three alleles, but as 'lyra' itself could be produced by more simply banded 'frenata' and 'zonata' the true pattern was that alleles of two different genes were interacting to give multiple combinations.

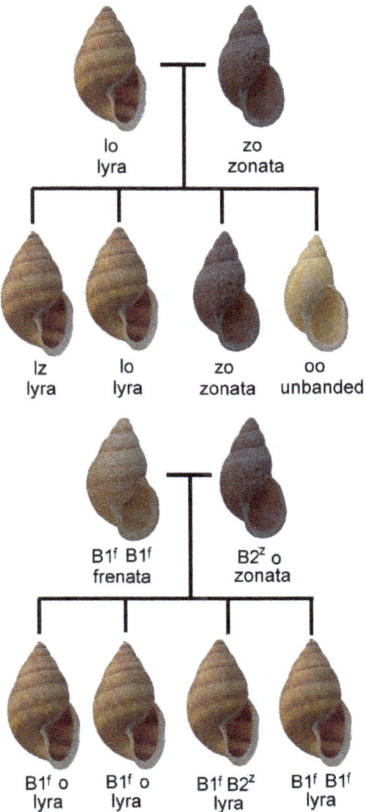

Fig. 3. Banding inheritance in *Partula taeniata:* basic form with 'lyra' (allele l) is dominant to all others, 'zonata' (allele z) dominant to unbanded (allele o).

As the genes for banding, lip and shell colour, and others for spire colour, interacted with each other to produce the shell pattern they could be considered as a single entity, a 'supergene'. A supergene was also present in *suturalis*, but here the situation was even more complicated due to the greater variety of colour forms and the genetic interactions which produce them. In shell colour, of the six alleles that were identified in *suturalis*, two showed only partial dominance, where heterozygotes have their own characteristics in some combinations; one allele was dominant to all others, but neither of these recessives were dominant to each other. There were also completely dominant and completely recessive alleles. All of this complexity leads to the extraordinary diversity of patterns found in *suturalis*.

Controlling the crosses between snails was not always easy. They found that at least some of the snails were able to fertilise themselves in the absence of suitable mates. Young adults were frequent self-fertilisers, accounting for some 10% of births in *suturalis* and 20% in *taeniata*, *olympia* and *aurantia*. They were not highly successful though as 93% of eggs produced this way failed to hatch. On the other hand, self-fertilisation seems to be the normal way of reproducing in *gibba* from the Marianas islands. In species that are unlikely to find a mate easily, being able to use a do-it-yourself method is highly advantageous. For this reason self-fertilising or cloning species are good colonists, as in *gibba*. This would have been beneficial when Moorea was first colonised by *Partula*, but over time became less and less important. In the early 1960s *Partula* could be found at extraordinary densities; as many as 20 per square metre. These were highly successful animals, at least in those days. Even though they only gave birth to a single baby after a gestation of 19 days *P. taeniata* could reach maturity in as little as 4 months and could live for 17 years, so one snail could theoretically produce up to 312 young in its lifetime, although the normal number is likely to be much lower.

Crampton had found no relationship between variation and environment and this led him to conclude that what he was looking at were what Darwin had called 'indifferent characters'. That is, they were not selected for or against, and so only new mutations and changes in migration patterns could change their frequency. Crampton thought that selection only acted to eliminate deleterious mutations, and that 'contemporaneous organic differentiation' could be observed by comparing his collections with those of Garrett or even within his series. Where changes occurred they must be due to the appearance and spread of new forms. The Professors disagreed; they were of the opinion that Crampton's claims that the frequency of different colour patterns had changed over the 50 or so years between Garrett's collections and his own were due simply to sampling differences. Specifically, they concluded that Crampton had penetrated further up some valleys than Garrett had done. They also concuded that chirality acts as a barrier to gene flow, so limiting the spread of any new forms, and that some sort of selection was keeping the abundance of different colour and banding forms steady.

The Three Professors paid particular attention to the old question of shell coiling. This had caught the attention of everyone who came across them since Bruguière's *Bulimus otaheitanus* and *Helix perversa*. As with most snails, *Partula* are basically dextral. However, three islands show common exceptions to this: Saiapan (where *gibba* is sinistral), Tahiti (*cytherea, nodosa* and *otaheitana* are normally sinistral) and Moorea (sinistral *olympia, tohiveana, mooreana* and mixed *suturalis*). *Samoana* is more constrained, with only three species showing any sinistrality (*canalis, conica* and *stevensonia*).

Fig 4. Just some of the banding varieties of *suturalis* figured in Crampton's Moorean monograph

With the discovery of the 'maternal effect' in delaying inheritance of shell coiling by a generation in the 1920s, the inheritance of coiling had been explained for *Lymnaea peregra*. In the 1960s it was not known if this explanation worked for other snails, or was just a special case in this species. In *Partula*, Murray and Clarke found that crossing dextral *mirabilis* with sinistral *tohiveana* resulted in dextral hybrids after a delay of a generation. This confirmed the maternal effect of shell-coil and showed that dextrality was dominant. They attributed coiling to a gene which was inherited in a straightforwardly Mendelian fashion, if delayed a generation by the maternal effect.

Combined with an understanding of snail embryonic development, worked out by none other than Henry Crampton back in 1894, it was now possible to see exactly what happens to the developing snail. As with all animals, the snail starts out as a single cell, dividing into multiple cells in a predictable way: from one cell to two, then from two to four, and so on. At the third division, when the four cells are about to divide into eight cells, something changes. Instead of producing eight equal sized cells, the snail embryo produces four large cells and four small cells. The small cells then rotate to the right or left, slipping into the grooves between the large cells, producing a compact ball of cells. This sounds odd but these 'spiral' embryos are completely normal for most invertebrates. Whether the small cells rotate to the left or the right is determined by the proteins present in the egg (proteins laid down by the mother before the egg was fertilised).

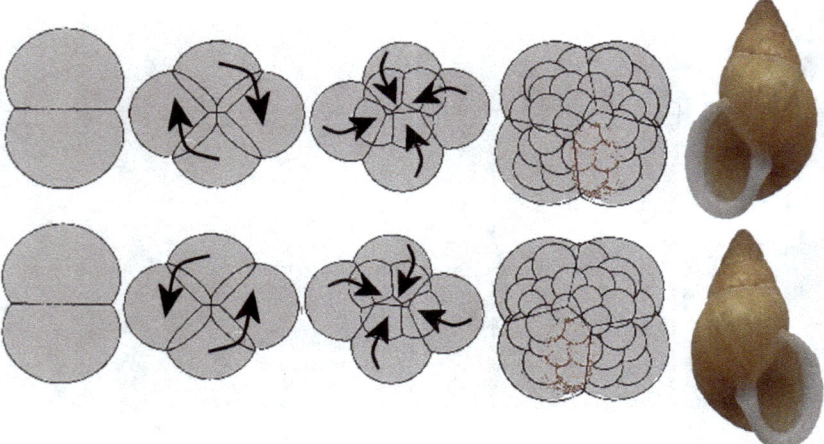

Fig. 5. Snail embryo development and chirality showing early stages of cell division, from (left to right) the 2-cell, 4-cell, 8-cell and 64-cell stages, with adult shells shown at far right. Top row – sinistral development, bottom row – dextral development. Note development is the same but in mirror-image.

This asymmetry of development leads to some genes being expressed more on the right side of the body than the left in dextral snails and vice versa in sinistrals. These in turn cause a gene called 'dpp' to make the mantle and the shell gland expand on that side. This makes the shell grow over from the right side to the left (in a dextral form), giving rise to the spiral or coiled shell typical of most snails. This is only part of the story though. As the shells starts to develop the strangest event in a bizarre embryonic development occurs. The embryonic snail with a newly developed small shell starts to convulse. The 'nodal' and 'Pitx' genes that are expressed mainly in the right side of a dextral snail and the left of a sinistral result in one set of foot muscles becoming more developed in one side than the other. These muscles run from the developing foot and the head to the shell. When these muscles start to contract they pull the shell round towards the head every 30 seconds. The more developed, stronger muscles cause the shell to turn in one particular direction. After minutes or hours of this turning, the shell suddenly twists through an angle of 90°. This torsion and differential growth results in the posterior part of the shell now being positioned over the head of the snail. This strange process results in a spiral snail shell and a strangely twisted nervous system. The evolutionary reasons for this 'torsion' are not clear.

In 1987 Johnson found that there was something more to the spiralling problem. He realised that sinistrals and dextrals were not simple mirror images of one another. This perhaps contributes to their feeling of 'wrongness' – a reversed photograph of a sinistral snail does not make it completely normal. In *suturalis* sinistrals are stouter than dextrals.

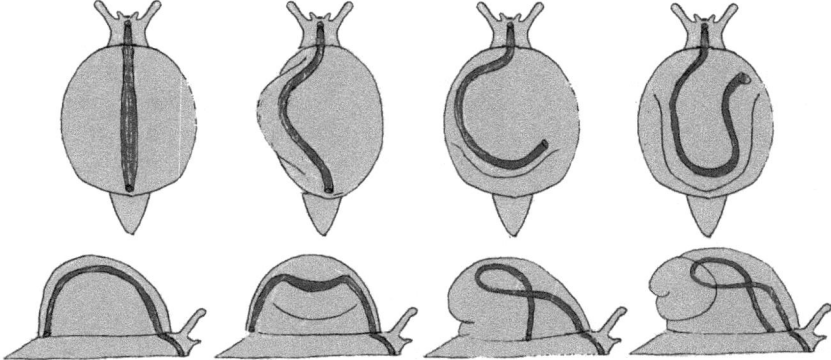

Fig. 6. Snail torsion. Snail outlines showing orientation of the developing shell and postilion of the gut shaded dark grey. Development stages are shown from left to right: early symmetrical development; the right side developing faster than left due to dpp gene expression; full torsion where the shell is pulled round through a further 90°; and continued growth.

The pattern of inheritance of coil direction is not merely of academic interest; there does seem to be a significant evolutionary aspect to it as sinistral and dextral snails do find it difficult to mate successfully. It is not impossible, but mating success is greatly reduced (an 80% reduction) so species that spiral in different directions are much less likely to hybridise than same coiling ones. A student of Jim Murray's, Carol Lipton, showed why this is the case in a detailed description of the courtship of *suturalis* and *taeniata*. Both species apparently have distinctive courtship behaviours: in *taeniata* the snails would chase one another in a slow motion 'dance', which seems unusual in a hermaphrodite animals, able to act both as male and female. After the dance the snail acting as the male would climb onto the other's shell. There it would crawl from one end of the shell in distinctive repeated figure of eight patterns. It before reaching its head down to the other and starting to probe for the genital opening with its erect penis. Once mating was complete they would reverse their roles immediately. Courtship in *suturalis* is much less romantic – they do not bother with the dance, the 'male' climbs on the shell of the 'female' and crawls from one end of the shell to the other repeatedly without the complexity of the figure of eight. After mating they would delay the reciprocity. Snails of this species are typically sinistral but can also be dextral. Sinistral and dextral snails will court one another but when it comes to seeking the genital opening a dextral 'male' will search for it on the right side of the 'female's' body instead of the left. This 'male' will only mate successfully if the 'female' twists her body round so that they are aligned, which does not always work.

Where *suturalis* coexists with the sinistral *mooreana, olympia* and *tohiveana* it is dextral. This has been interpreted as character displacement: the presence of sinistral competitors effectively forces *suturalis* to switch to a dextral form. In reality the dextral individuals are at an advantage over the sinistral ones as they produce a higher proportion of viable offspring. Why this might be remains a complete mystery. The dextral species *taeniata, exigua* and *miriablis* do not cause such displacement, possibly because they are all smaller than *suturalis* whereas the sympatric sinistral species, and the dextral *aurantia* are all at least as large as *suturalis*. This makes mating attempts between them quite likely, so size and coiling create barriers to hybridisation. There is no barrier between *suturalis* and *aurantia* and they do seem to have hybridised freely.

Where mixed coil populations occurred their success would be relatively low, so they would be selected against and one form would come to dominate the other. In accordance with this there was only a narrow zone where mixed coiled snails occurred. In the main populations the rare form would produce fewer young because it tended to mate with the commoner opposite coil; as the selective advantage favour the commonest form, they are kept separate.

Shell spiralling is therefore of great importance for the evolution of separate

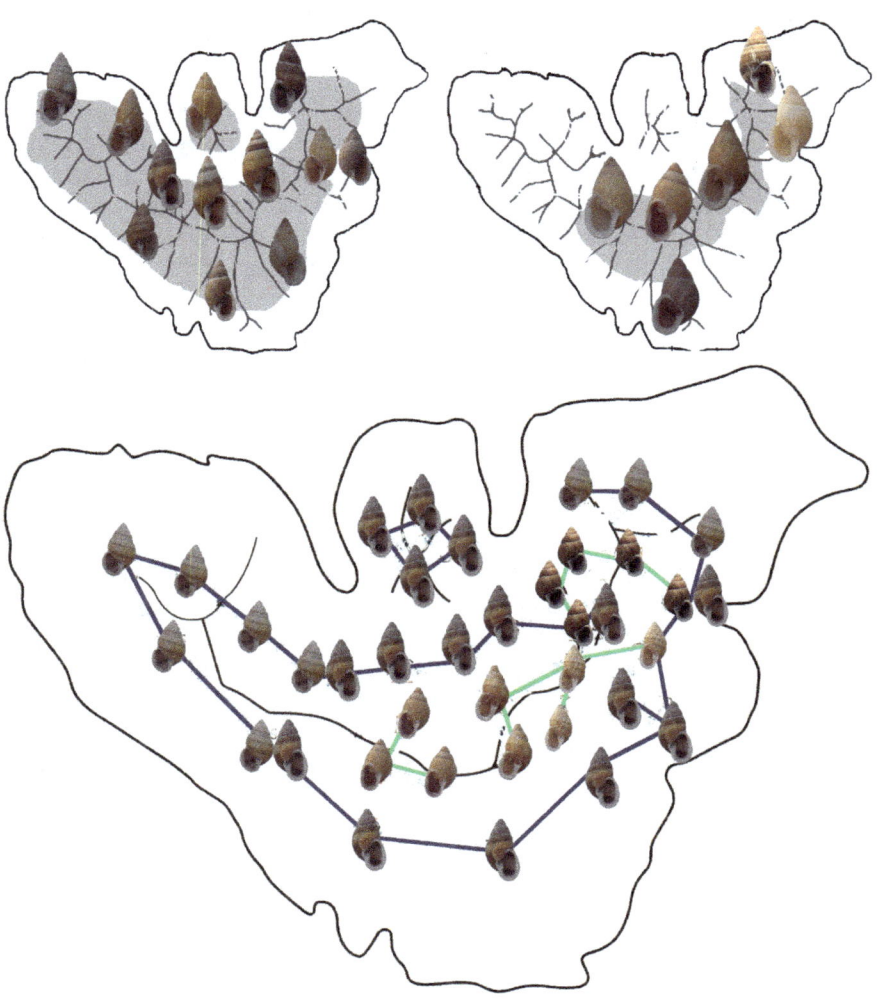

Fig. 7. *Partula suturalis* chirality on Moorea. Top left: *suturalis* range and chirality, showing sinistral forms predominating except in the centre of the south. Top right: distribution and chirality of species related to *suturalis*. Bottom: pattern of gene flow between *suturalis* and related species, *suturalis* populations are linked by black lines, other species by grey lines (figure modified from Barton *et al.* 2007).

species and keeping them distinct. In order to understand the evolutionary history of the species, coiling needs to be understood.

The first attempt to determine the evolutionary origins of *Partula* species, other than Crampton's speculations, was by Bailey in 1956. Bailey's significance to science is as the breeder of special strains of mice that have been vital in cancer research, but the early days of his research career he dabbled briefly with *Partula*.

He used Crampton's shell measurements to calculate what were called 'phylogenetic distances'. The mathematical techniques for this were fairly complex and can be seen as a step forward in genetics research, but were quickly superseded. Not surprisingly, Bailey's paper on *Partula* has received very little attention. Research quickly moved on to looking at the genetic material itself. However, *Partula* chromosomes turn out to look very similar in all species. The appearance of chromosomes is not particularly useful though, as the genetic material that matters is too small to be visible, so the actual chemistry itself needs to be examined.

Although the structure of DNA was worked out by Francis Crick and James Watson in 1953 and gene sequencing now seems commonplace, it was not practical to isolate pieces of DNA until the 1970s. Before then it was only possible to identify the proteins that the DNA produced. So when the first attempt to work out the genetic evolution of *Partula* were made they were restricted to differences in enzymes. The proteins that make up slightly different forms of enzymes ('alloenzymes' or 'allozymes') were identified by electrophoresis. Protein electrophoresis analyses the proteins in a sample by length, separating these molecules by applying an electric field to a gel containing the sample. This moves negatively charged molecules through the gel; the shorter molecules move faster than the large ones and are transported further. This produces a pattern of bands of different proteins in the gel, which can then be compared between the species being examined. The technique was first used in 1963 and applied to *Partula* in 1969 by Schwabl and Murray. This study compared populations of *suturalis, dendroica, aurantia* and *tohiveana*, confirming that *suturalis* and *dendroica* were indistinguishable and also that some *suturalis* and *aurantia* specimens were hybrids. As the techniques developed it became possible to undertake more detailed studies; a study of 20 enzymes and their associated alleles by Johnson, Murray and Clarke in 1977 found no significant differences between the levels of variation found within and between species, meaning that the species were no more different from one another than were the different populations. In 1986 they came to much the same conclusion again: there was high variability within species but little divergence between them. This may have been due to the species having separated relatively recently. Some populations of *taeniata* in the south of Moorea differed from other populations more than the differences between species elsewhere on the island. There seemed to be no correlation between allozymes, morphology and isolation. The allozyme study identified two species complexes: the *taeniata* complex (*taeniata* and *exigua*) and the *suturalis* complex (*suturalis, aurantia, tohiveana* and *mooreana*). *mirabilis* formed a link between the two groups, hybridising with *taeniata* in the wild and with *aurantia* and *tohiveana* in the laboratory.

A larger study in 1996 looked at 19 variable enzymes. This found that populations of *taeniata* and *suturalis* that coexisted were more different in proteins in the south than in the north. In the south these two species were very different from one another, but less so in the north. In addition, southern and northern *taeniata* showed a similarly marked difference, even though there was a complete range of intermediates between these populations. This relatively high level of distinction of the southern *taeniata* might be explained by invasion of genes from Tahiti, with two separate invasions, possibly by snails blown over from Tahiti by a typhoon. Using a measure of genetic distance they constructed an evolutionary tree that showed that the extreme southern *taeniata* are more closely related to Tahitian species than to any Moorean population. The southern *suturalis* did not seem to have this special history but it was speculated that as this is a larger and more mobile species than *taeniata* it may have had more gene flow, obscuring

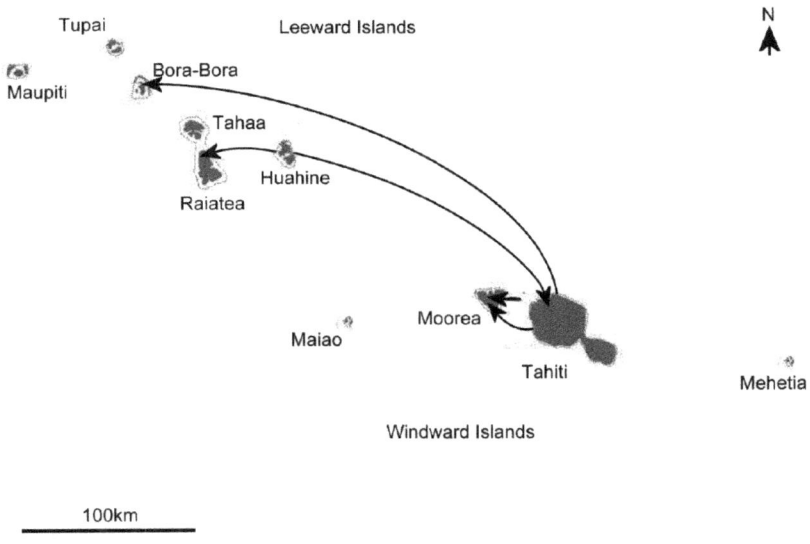

Fig. 8. The 1996 hypothesis of island colonisation showing initial colonisation of Huahine and then dispersal out to Raiatea and Tahiti, followed by dispersal from Tahiti to Moorea and Bora Bora

its ancestry. This 'molecular leakage' between populations and species confuses the reconstruction of their evolutionary history. In a wider perspective, samples from other islands indicated that the Huahine island species were the earliest to evolve, followed by those on Raiatea. The single species on Bora Bora grouped with the Moorean and Tahitian species. This more or less fits with the general speculation that the older, western islands may have supported *Partula* before the snails arrived on Moorea and Tahiti, but does not fit a geographical or age pattern precisely. As the first attempt to reconstruct the pattern of island colonisation it was not clear if this was accurate: more studies were needed.

Although proteins remained the basis of the main studies of the relationships of the *Partula* species into the 1990s, gene sequencing was becoming available as a practical technique. This was developed back in 1972 but for many years remained too difficult and costly for widespread application. It was not until 1991 that it started to be applied to *Partula*. It was not the sequencing that seems so commonplace today, but 'restriction fragment length polymorphisms' (RFLP), a variant on the electrophoresis that was now well established for proteins. The DNA could be extracted from a cell and cut into pieces using enzymes which broke the DNA strands at specific genetic sequences. These fragments were then separated by size through electrophoresis as before. They were then matched up to a reference genome, that of the standard laboratory animal, the clawed toad *Xenopus laevis*, to create a 'restriction map' of the mitochondrial genome for each snail. These were then classified into categories, or genotypes, depending on whether they had a particular fragment or not. Samples could then be grouped statistically, based on how similar their maps were.

The outcome of this laborious process was that 32 genotypes were identified in 14 species. 12 populations were polymorphic in that they showed at least two different genotypes, with one population having as many as four variants. The results were similar to earlier studies: on Moorea the most variable species were *suturalis* and *taeniata*, and some species, such as *taeniata* and *exigua* showed no clear separation from one another. This study extended beyond Moorea, with a few samples from other islands and found that some genotypes were found across almost all species on Moorea and also in some Tahitian species, supporting earlier ideas that there may have been movement of species between the two islands, *suturalis* on Moorea seemed to be related to *nodosa* on Tahiti, and southern populations of *taeniata* had what looked like Tahitian genes.

Sorting out the evolutionary history of the species within the islands has seemed like a thankless task. First of all it is confounded by environmental influences; 70 years after Crampton collected his Tahitian data Ken Emberton re-analysed the measurements and found that *otaheitana* were more elongate, had a higher frequency of banded forms and simpler-shaped mouths in wetter, more shaded valleys than in the dryer, more exposed ones. On Moorea the pattern was less clear, but the now notoriously variable *suturalis* was longer at higher, wetter sites, being reversed from the *otaheitana* pattern. This means that the fact that snails in one valley may be clearly distinct from their relatives in the next valley may not be because they are different subspecies, evolving in different ways, but simply because one valley is wetter, or more sheltered than another.

Even if the different populations are evolving in different ways their adaptations are confused by extensive hybridisation. Despite Clarke, Murray and Johnson's conclusion that shell spiralling acted as a barrier to hybridisation, some hybridisation did seem to be occurring. Sinistral populations of *suturalis* did not interbreed with the dextral *aurantia* but in some valleys there were both dextral and sinistral *suturalis*, and in these it was impossible to identify clear *aurantia*; there were only sinistral *suturalis* and dextral hybrids. Where dextral *suturalis* lived alongside the sinistral species *mooreana, tohiveana* and *olympia* hybridisation did not occur, but this was not the case in the area where *suturalis* changed from dextral to sinistral. Murray and Clarke concluded that *suturalis* was basically a sinistral species and that the dextral populations were the result of character displacement. Where all the snails were sinistral, the occasional dextral *suturalis* would be at a disadvantage in that it would find it difficult to find a mate. It would however, have an advantage in that it would be much less likely to mate with the wrong species. If this risk was great enough, then the dextral form would become the commonest one in valleys occupied by the sinistral *mooreana*. All of this hybridisation though raised a question: were the different forms named by Pease and Crampton really different species?

Chapter 6. The meaning of species

Partula clara incrassa

In determining that Moorean *Partula* could hybridise quite widely Clarke, Murray and Johnson called into question the meaning of species in these snails. Essentially, if they could hybridise were they really distinct species? The easy answer would be no, but then how could such clearly distinct forms as *suturalis* and *tohiveana* be called the same species?

When Garrett had first collected in the Society Islands in 1860 there was no great difficulty in recognising species. Prior to the publication of Darwin's "*Origin of Species*" in 1859 almost everyone had accepted that they were God's immutable creations. This meant that each was a fixed, recognisable entity. If some hybridised that was merely a curiosity and did not call into question the validity of the species themselves. I have no idea what Garrett thought a species was, but as he was collecting only a year after the revolutionary book was published it is unlikely that he gave much thought to the possibility of evolution then. He did note that hybridisation was widespread in some valleys, but just identified specimens as hybrids broadly, without explaining his reasons for this attribution. No-one attempted to test hybridisation in *Partula* for another 100 years.

Although the theory of evolution by natural selection was only 67 years old when Crampton started collecting, it was already firmly established. Even so, there was no discussion as to what a species was, even though it was no longer thought to be immutable. Species concepts had moved on in only one regard, that was the formalisation of units at a lower level: subspecies. Additional names for some forms had been in use almost since the start of taxonomy, in the form of varieties. Pease had named many varieties from Garrett's specimens. Then in the late 1800s these started to be called subspecies.

Using a third name after the species to denote a subspecies, without the use of 'var.' to denote a mere variety, was first practised by the German ornithologist Herman Schlegel in 1844. At the time this had little, or no, evolutionary meaning. This was largely overlooked, but was followed by John Cassin 10 years later who introduced this use of trinomials into American ornithology. There it was picked up enthusiastically by Robert Ridgway, who justified the approach in 1879 and then boldly dropped the term variety in his 1881 summary of North American bird nomenclature, creating trinomial names from all the previously existing varieties. In Europe there was considerable resistance to this approach, although a meeting was held at the Natural History Museum in London in 1884 to discuss the use of trinomials. This gave the idea only limited support; the ornithologists at the meeting thought it might be practical in some situations but those interested in other groups generally gave it a negative reception. Even so some European ornithologists started to follow the American lead, especially in Germany. One of these was Ernst Hartert who started using trinomials in 1887. Hartert was the curator of Lord Walter Rothschild's bird collection, and it was this proximity to Rothschild that led to the establishment of the subspecies in biology.

The 18th and 19th centuries had been the era of gentleman naturalists and eccentric collectors. While Charles Darwin's death in 1882 effectively marked the transition from the gentleman to the professional scientist, Rothschild at the turn of the 20th century was the apogee of the collector. The greatest collection of animal specimens ever amassed by one man was accompanied by scientific interest in diversity and taxonomy. This obsession of Lord Rothschild's overrode almost everything else. His specimens are now housed in the Natural History Museum's property at Tring. Sadly though, the collection is not the memorial it should have been, for during his life obsession got the better of sense and parts of it had to be sold. At the time the Natural History Museum did not appreciate its value, letting significant parts of it pass them by, in particular losing out on the opportunity to secure the astounding bird collection that Hartert had curated. Before it started to be broken up though, Rothschild published many important papers along with Hartert and his entomologist Carl Jordan. Jordan was the leading academic influence in Rothschild's group and an expert on systematic practice as well as on taxonomy, and his support for Hartert's use of trinomials for the bird collection led to Rothschild and Jordan's use of the concept in 1895 in major works on moths. From 1895 Rothschild, Jordan and Hartert supported the proposition that naming subspecies would be useful, and that subspecies represented distinct forms that could evolve into full species in time. As a result of their work subspecies quickly became accepted categories in all animals, not just birds.

Rothschild's bird collection was so great that it required more than just Hartert to manage it. One of the additional curators was another German taxonomist, Ernst Mayr. Mayr worked with Hartert in England and from 1932 was a curator at the American Museum of Natural History which had bought Rothschild's 280,000 specimens. Mayr continued to use the Rothschild approach to taxonomy and started to develop an interest in phrasing the taxonomic concepts of species and subspecies in evolutionary terms. This necessitated a return to a largely overlooked problem – what exactly was a species?

Although species are a familiar concept they are actually very difficult to define scientifically. The great French biologist George Cuvier tried to define species as early as 1798:

> "To accept two creatures, which differ more or less, as being only varieties of one species, it is necessary firstly that the distinguishing characters belong to those which are known to vary; secondly that there should be cause for variation; and thirdly that they should produce fertile offspring when crossed. Therefore two wild forms which live at the same place in the same climate, without interbreeding, and always maintain their differences, have to be regarded as different species, no matter how trifling the difference might be" (original text in French).

Darwin did not find Cuvier's approach fully satisfactory; he wrote in the "*Origin of Species*":

> "No one definition has satisfied all naturalists; yet every naturalist knows vaguely what he means when he speaks of a species."

Here he was pointing to the arbitrary nature of the 'typological species concept' whereby a species was a group of organisms that shared some characteristic, but who could decide on what characteristics mattered? This was refined slightly by the 'recognition species' concept. Under this, species were groups that recognised themselves as such by some means such as mating behaviour. This was made more meaningful by 'mate-recognition species' whereby the recognition was for mating. This would mean that species were defined on reproductively relevant characters (as Linnaeus had proposed for plants).

In 1940 Ernst Mayr combined Cuvier's definition with evolutionary theory to come up with a definition that remains the best devised to date. This is the 'biological species concept': a species is a group of actually or potentially interbreeding natural populations, which are reproductively isolated from other such groups. This is a practical concept: individuals belong in different species if they cannot hybridise if placed together. However, there are many exceptions to this: many species which can, and indeed do, hybridise but which we wouldn't place in the same species. Horses and donkeys are the classic example; they hybridise to produce mules. Reproductive isolation therefore must be more than just inability to hybridise; it could be argued that horses and donkeys are still species because their hybrid mules are sterile. Except mules are not always sterile; very rarely a female mule will give birth to a perfectly viable foal. They are not truly sterile, they just have extremely low fertility. So the boundary can be shifted further, reproductive isolation is not the inability to hybridise, nor is it the inability to produce fertile hybrids, rather it is the inability to produce fully fertile hybrids. But now where do we put the line – exactly what level of fertility do we require? Our attempt to make a rigorous definition has slid back into subjectivity. This has caused the biological species concept to fall into disrepute – it is just not rigorous enough.

There have been various rephrasings of the reproductive concepts: the 'competition species concept' sees the species as the group among which reproductive competition occurs – i.e. those that can interbreed. This does not change the problem. More constructively there have been many attempted modifications of the biological species concept. The 'evolutionary species concept' takes a different route: a species is a lineage which is evolving separately from other lineages, with its own evolutionary tendencies. This seems to be a statement of the obvious but does not help us to define a species in any practical way. It was further tied into reconstructing evolutionary history by the 'phylogenetic' or 'cladistic species concept' where the lineage is defined as descending from a branching point in an evolutionary tree. What these two versions deal with are Evolutionarily Significant Units; populations which may have a different evolutionary future from others, but are these actually species? The answer has to be: not necessarily. ESUs simply have the potential to be different, it does not mean that they are different now; they may be species or may become species in the future. If we want to define a species now, this does not help at all.

Taking a completely different line the 'ecological species concept' sees a species as organisms that use the same niche that is minimally different from other such groups. This concept has never caught on; it differs too much from traditional systems and relies on the ecological 'niche', another arbitrary concept.

The most rigorous definition is one that is widely used in phylogenetic studies. There is a current tendency to define species based purely on DNA. These genetic species sound as if they are highly rigorous – each species differs from its relatives by the same proportion of its DNA after all. However, this is simply an arbitrary division and returns us to the typological concept that we started with: a species is different because we consider it to be different, based on a particular character (in this case the sequence of a gene). For all the efforts to define and redefine species progress has been almost non-existent. The biological species concept remains the most practical, and the most widely used. It is valid so long as it is recognised that speciation can involve the gradual transition from freely interbreeding populations (conspecific), through increasingly isolated forms that can hybridise, to incipient species with only partial fertility, and finally to fully reproductively isolated species.

Returning to *Partula*, Murray and Clarke noted the problem posed by the hybrids on Moorea. They were of the opinion that they could still be recognised as species: "Despite the equivocal nature of some of these taxa we have continued to use the traditional names, for several reasons. First, each taxon corresponds to an entity that can be discriminated on the grounds of morphology and distribution. Secondly, the use of familiar names makes it easy to compare our material with Crampton's.

Thirdly, there seems to be no reasonable way to accommodate the group within a conventional Linnaean scheme. It would be easy to argue for as few as four or as many as ten 'true' species." They thought that Society Island *Partula* were evolving separately on each island and that within each island the different forms were in the process of diverging; speciation was still in progress. This was a combination of allopatric and parapatric speciation: allopatric being the divergence of physically separate populations (on islands or certain valleys) and parapatric being the slower diverging of only partially separate populations.

Whether the different recognisable *Partula* forms are separate species, or populations in the process of speciation, or a mixture of both, their evolutionary origins should still be explicable. The electrophoresis studies of proteins gave results of varying reliability. The first study to include a wide range of species from different islands, including from the western Pacific was a study of allozymes by Johnson, Murray and Clarke in 1986. This seemed to show that the Raiatean *turgida* had been the earliest *Partula* to evolve. The next most basal species was *gibba* from Saipan in the far distant west Pacific. This was followed by the species on Huahine, which then gave rise successively to the rest of the Raiatean species, the odd pairing of *lutea* of Bora Bora and *radiolata* of Rarotonga, and the Moorean and Tahitian species, with Tahiti apparently colonised from Moorea. Their allozyme study of 1996 gave similar results, although it did not include *turgida* this time. Making sense of this pattern would require that the western Pacific islands were colonised from the Society islands twice, giving rise to *gibba* and *assimilis* separately. The pattern within the Society Islands is plausible, with colonisation from the old western island to the younger eastern ones, possibly with movement backwards and forwards between Raiatea and Huahine.

In 1996 Bryan Clarke directed his lab towards modern DNA sequencing, taking on Sara Goodacre as a PhD student, with Chris Wade as a postdoctoral researcher. It was now possible to determine the sequence of the DNA code directly, rather than looking at the proteins it produced. The DNA extracted from snail tissue was chopped into sections using enzymes and markers that identified specific sequences. This allows particular genes to be sequenced and compared. In the 1970s and 1980s the gene would then have been subjected to a very laborious process of a multitude of chromatography tests to identify each individual chemical base at each point along the gene. Now it was possible to place the selected gene in an automatic sequencer. These techniques identified *Eua* as the earliest genus in the family, just as the anatomy had indicated to Kondo back in 1955. Goodacre and Wade concluded that partulids probably originated from a fragment of an ancient land mass far to the west: what has been called the 'Eua continental fragment'. This broke off from New Caledonia and drifted eastwards, sinking and fragmenting into the small island of Eua

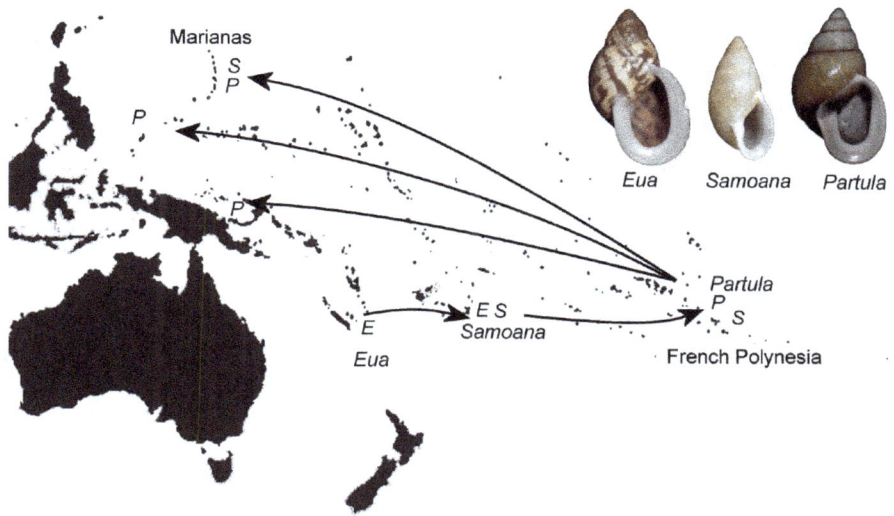

Fig. 1. Origins of the partulid genera *Eua* (E), *Partula* (P) and *Samoana* (S)

around 6 million years ago. This is inhabited by the oldest partulid genus, not surprisingly called *Eua*. The snails that remained on Eua and the nearby islands have changed little, but some dispersed eastwards to the Samoan islands and evolved into *Samoana*, a genus which continued dispersing, finally reaching the Society Islands in the centre of the south Pacific. *Partula* seems to have had a similar history, also dispersing eastwards through the Samoas, to Polynesia, but it also seems to have managed the return journey, drifting north-westwards to the Marianas.

Within the Society Islands Goodacre and Wade's studies again gave the Raiatean *turgida* as the earliest species, although this time along with *hebe*, also from Raiatea. These were followed by two groups, one with the Huahine species, most Mooreans and one Tahitian species, while the other group contained most of the Tahitian species and one Moorean. Strangely this latter group placed some Moorean *suturalis* with Tahitian *nodosa* and, more oddly, included the Raiatean *tristis* with Tahitian *affinis*. If this was correct it seemed to show a great deal of movement between islands, not just the progressive colonisation from west to east, ending with Tahiti but several such colonisations.

It seems unlikely that snails could move easily between islands, but there is no other plausible explanation for their distribution on the volcanic islands scattered across the Pacific Ocean. They could cross the sea by drifting on floating wood, but that would expose them to sea-water which *Partula* would be unlikely to survive.

The other possibility would be dispersal by the wind. This may sound even more implausible but hypothetically small juveniles could be carried by storm winds. This is the most likely answer for species such as the *Samoana* species: *attenuata* was found on all the Society islands and *burchi* on Tahiti and Moorea but these are high altitude species that could not possibly raft across the sea and then slowly work their way up the mountains to their favoured cloud forest.

By 2002, unravelling the evolutionary history of *Partula* seemed to have gone as far as it could. Almost all the available samples had been studied at the protein level, mitochondrial DNA alleles and sequences. The only remaining approach would be to examine a much wider range of genes, but this seemed a low priority. However, in 2003 Jack Burch mentioned to an Associate Professor at the University of Michigan, Diarmid Ó Foighil (now the Director of the University's Museum of Zoology), that he had collected around 600 *Partula* back in 1970. These were the snails he had collected in collaboration with Yoshio Kondo. As they had been freeze-dried their DNA should have been preserved intact. Ó Foighil is a geneticist and he realised that these specimens could solve some of the mysteries of the evolution of the *Partula* species, despite the fact that most were now extinct. His research interests until this point had been on marine invertebrates, now he found himself drawn into the *Partula* world. He and Burch, along with Taewan Lee, a curator in the museum, embarked on a study of these specimens, along with samples from both captive and surviving wild populations.

The previous studies had not come to any very clear results, and the different types of genes that had been looked at had seemed to tell different stories. When undertaking genetic studies there are two main sources of DNA. One is the nucleus of the cell, the other the small mitochondria within the cells. These mitochondria are membrane wrapped structures that contain their own genetic material, separate from that of the cells in which they reside. The reactions that occur within the mitochondria are the main sources of energy for the cell. Within the cells these structures act as almost independent organisms, cloning themselves without mixing their DNA with that of the cell. This cloning means that mitochondrial DNA normally remains pure, whereas the DNA in the nucleus is a mixture of both maternal and paternal components. Furthermore, in every generation the mitochondria are passed on in the egg, and so are only inherited down the maternal line. This makes it relatively easy to reconstruct the evolution of mitochondrial genes: each genotype has been inherited maternally and differences between genotypes have arisen only by mutation. In contrast, nuclear genes are constantly being shuffled between paternal and maternal lineages, so their inheritance is far more complex. Where hybrids are concerned, a hybrid will have the mitochondrial type of its mother, but its nuclear genes will be an equal mixture of both parental species. As we know that hybridisation was common on Moorea at least, it is not surprising that early attempts to use nuclear genes in studying *Partula* failed to produce any useful patterns. Mitochondrial genes seemed to work better, and so the Michigan project started with mitochondria. The main part of Burch's collection was from Tahiti so these extensive samples were examined first. These included several species never before sampled and samples from many different populations.

The results from Tahiti were far from reassuring. Rather than falling neatly into different species, the samples divided into three main branches, containing a total of five groups (labelled 'clades', or branches, 1 to 5). Some of the species grouped quite reasonably: clade 5 contained only *clara* and *hyalina*, *filosa* was restricted to clade 2, *producta* and *nodosa* to clade 3, but most were scattered between the different clades. Of all the 'species' sampled, *otaheitana* individuals were found in almost all clades and *affinis* in three. This could mean that *otaheitana* and *affinis* actually contained many different species, or, more realistically, that there had been widespread hybridisation spreading these mitochondrial genes between species. So it would seem that *affinis* and *otaheitana* were able to hybridise whereas the other species were more separate. This was perhaps not very surprising as Crampton had originally described *affinis* as a subspecies of *otaheitana* and had noted how variable and widespread the latter was.

A second publication followed, adding the Moorean samples to the dataset. This found the earliest group from Moorea and Tahiti were three Tahiti clades containing *otaheitana, affinis, producta, filosa* and *nodosa*. These Tahitian species seem to have given rise to the Moorean ones, but Moorea in turn, was the origin of Tahiti clade 1 (more *otaheitana* and *affinis*) and 5 (*clara* and *hyalina*). It seems that a snail arrived on Tahiti from Huahine or Raiatea to the west, diversified into different groups there and at least one of its descendants made the relatively short crossing to Moorea. Murray, Clarke and Johnson had earlier speculated that Tahitian snails might have colonised Moorea but the new study suggested that Moorean snails must also have returned to Tahiti at least three times: the snails of the two islands were far less isolated than they may have appeared to be.

Of the Moorean species, there were two main groupings but again with jumbling of species between them, probably due to the extensive hybridisation that had been found in the earlier studies. There seemed to be some suggestion of some early geographical separation, with populations grouping into north and south, or east and west, forms before spreading and mixing.

One of the most unexpected results of the genetic studies was that one population of the Tahitian *clara* lies within one of the Moorean clades. This had been described by Crampton as *Partula clara incrassa*, a form very slightly different from other *clara*, but genetically this is not *clara* at all. Here there is evidence of a previously unrecognised species, a 'cryptic' species. These are often identified by genetic studies: species which have diverged from other species because of undetected biochemical adaptations or through isolation, but which have not changed physically. It may be impossible to identify any physical differences between these species and yet the DNA indicates that they are good species, perhaps incapable of cross-breeding. In the case of the Polynesian species *clara incrassa* looks like a slightly odd *clara* but is descended from Moorean *taeniata*. A second oddity in the *taeniata-clara-incrassa* grouping is *hyalina*. This is a close relative of *clara*'s; it is the same size, same general shape, has similar internal anatomy and, like *clara*, it is an unusual *Partula* in favouring relatively dry and open areas. Although they are similar they are very easy to tell apart: *clara* is brown whereas *hyalina* is white, it is a consistent difference and in some areas the two species can be found together. Genetically though they are completely indistinguishable, or at least that is what the mitochondrial DNA says. The earlier allozyme studies of proteins produced from nuclear genes could distinguish them, so are they the same? Morphology says they are closely related but separate; mitochondrial DNA says they are identical but allozymes suggest there is some sort of difference. At the moment no-one knows what these two are; they are thought to hybridise, but this has never been tested properly.

Partula hyalina is also odd in being the only *Partula* to be found on more than one island. It is found throughout Tahiti and also on the Cook islands. Genetics reveals that the snails in the Austral and Cook islands are descendants of those on Tahiti. This is evidence of recent movement, presumably human introduction, from Tahiti to the Australs and Cooks within the past 30,000 years.

The *Partula* DNA story did not quite stop there. In 2005 a new *Partula* was found on the very highest point of Raiatea. Being able to examine DNA of the new *Partula meyeri* prompted the sequencing of Burch's specimens from Raiatea. These were much more limited than the Moorean and Tahitian collections: even with the addition of samples from captive populations, only six species were available out of the 34 described from the island. This time Ó Foighil's group looked at a nuclear gene as well as a mitochondrial one. With fewer samples than the other studies, this produced a simpler pattern, but even so, there were problems. Mitochondrial genes placed *dentifera* and *tristis* next to a pairing of *turgida* and *faba*, whereas nuclear genes put them with *meyeri*. Either arrangement is in marked contrast to earlier attempts to reconstruct their evolution which put *turgida* at the base of all *Partula*, far from *faba*. This highlights an important point; it is unwise to rely on any one source of data: different genes may tell different stories. The fact that both mitochondrial and nuclear genes agree on *turgida* and *faba* going together makes it seem probable that the earlier placement of *turgida* was wrong.

The story of the evolution of the *Partula* species is still far from clear. At present, the various studies seem to show that Society Island *Partula* originated in the west of the archipelago; the oldest populations are probably those of Huahine, which colonised Raiatea and Bora Bora and latterly Tahiti. Moorea was then colonised from Tahiti and some of the resulting species on Moorea recolonised Tahiti. Within the islands there has been a great deal of hybridisation on Moorea with genes flowing between all species, even if by indirect routes. On Tahiti the evidence is less strong but there has been at least some hybridisation.

Even if the full picture of evolutionary relationships remains obscure, the molecular studies have been able to show up other interesting aspects of the evolution of these snails. In 2001 and 2002 Sara Goodacre published two studies of nuclear DNA genes covering 475 samples of 11 species from Tahiti, Moorea and Huahine. As Moorea had been the focus of Bryan Clarke's sampling the majority of specimens were from that island (365) but the samples were not evenly spread between the species, ranging from 268 samples of *taeniata* to just one each of *mirabilis* and *tohiveana*. On Moorea there was little sharing of genetic forms between valleys except in *mooreana* and north-western *taeniata* where genes spread over two adjacent valleys. The most widespread genes and the greatest diversity (0.44

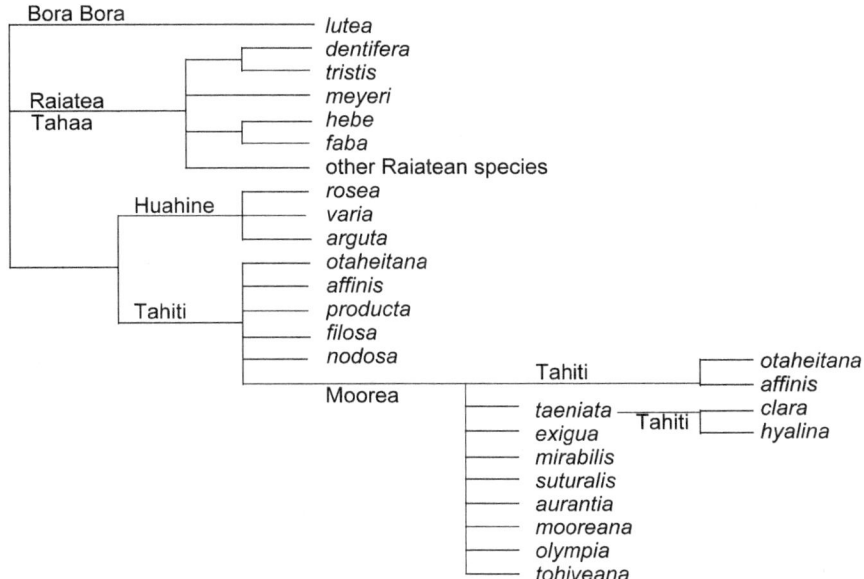

Fig. 2. The *Partula* phylogeny as currently understood with a plausible pattern of island colonisation. Note that only a few species from Tahiti, Raiatea and Tahaa have been studied.

variants or 'haplotypes' per sample) were in *suturalis*, which fits with this being the species most capable of hybridising with the others. The smaller sample of Tahitian genes were all site specific, again the species with the greatest hybridisation ability, *otaheitana*, was the most diverse (0.48 haplotypes per sample). The other species had less diversity, for example, the Moorean *taeniata* had only 0.14 haplotypes per sample. In contrast *varia* of Huahine was remarkably lacking in diversity (0.04). This species had just two haplotypes, one of which was spread across the island, the other restricted to the island's northern edge. This suggests that *varia* populations had been dramatically reduced in the recent past but had then expanded rapidly. Passing through this population 'bottleneck' had given rise to a widespread species but little genetic variation.

Even though its diversity was only moderate, Sara Goodacre found *taeniata* to have some areas of interest. One morphologically variable population had more genetic variation than the rest of the species. This population came from a site where two different physical forms of *suturalis* coincided, suggesting that for both species this was the border between two areas with notably different selection pressures. The alternative possibility, that it just happened to be the area where different subspecies overlapped for both species was unlikely as it was a very small area and the subspecies genes should have been able to spread much further. This gene flow would have blurred any random local changes (genetic drift) but may have not been enough to prevent localised responses to selection pressures. Working against the idea of local selection was the absence of any obvious ecological differences in the area. For example one form occurred in both the lowland forest and mountain scrub of the very wet Urufara ridge and also in the exceptionally dry Aareo and Vaitapi valleys. So it would seem that if there was a selection pressure it was not adaptation to habitat or climate. Goodacre suggested that this pressure may have been something historical, such as the presence of a bird that acted as a selection pressure on banding pattern. This was the 'apostatic selection' proposed by Bryan Clarke in 1962. Unfortunately for this idea, there is no suitable bird that could have been a *Partula* predator. The closest to a possibility is one of the two kingfishers of the islands but these are not known to be significant snail predators. The bird-life of Polynesia today is somewhat different to what was present in the past; across the Pacific the arrival of Polynesian islanders around 1,000 years ago was followed by the extinction of many bird species which today are known only from bones found as subfossils or in archaeological waste heaps. In the Society Islands, there are bones of an extinct starling. *Aplonis diluvalis* is known from a single bone from the archaeological site of Fa'ahia on Huahine island. A second possibility was the Raiatea thrush painted by Georg Forster in 1774 on Raiatea. Intriguingly, just as Cook's first voyage to Raiatea resulted in the collection of the first *Partula* specimen, which was then painted, named and lost, so the third voyage led to a new bird being collected, painted and then apparently passed on to Joseph Banks. Just as Gmelin rescued the name *faba* for the snail by copying Martyn, so Gmelin coined the name for this bird, accepting the Forsters' tentative identification as a thrush and adding the old name for Raiatea, making it *Turdus ulietensis*. Just like *Partula faba*, the *Turdus ulietensis* specimen then vanished and all that is left of the species is Forster's painting. This is clearly not of a thrush (*Turdus*) and looks most like one of the glossy starlings. It is now called the 'bay starling' or Raiatea starling, *Aplonis ulitiensis* and was probably a close relative of the Huahine starling, if not the same species. Glossy starlings are known to feed on snails at times and there is a record of the Asian glossy starling *Aplonis panayensis* systematically hunting for snails. It seems highly probable that the Society Island glossy starlings fed on snails some of the time, and so it may well have been the case that selection of different banding forms was created by predation, in just the way that thrushes select banded snails.

Fig. 3. Georg Forster's Raiatea starling. a) Forster's original 1774 painting of '*Turdus badius*' based on the real bird, original in the Natural History Museum, London.; b) Keuleman's 1880 version of '*Merula ulietensis*' based on Forster's description and making it look more like a thrush.

Fig. 4. Jim Murray and Bryan Clarke working on their 1998 study of *Samaona* species from the Marquesas islands. This was the last paper by the Three Professors. Photo: M. Johnson.

In 2014 Bryan Clarke died at the age of 81. Of his many contributions to science, newspaper obituaries drew attention to two specific elements of his research: firstly to the role of predators in maintaining genetic polymorphisms through apostatic selection and secondly to the less positive impact of predators on *Partula*. This latter aspect became the focus for *Partula* research from the last quarter of the 20th century when all previous phases of *Partula* research came to a catastrophic close.

Chapter 7. The alien invasion

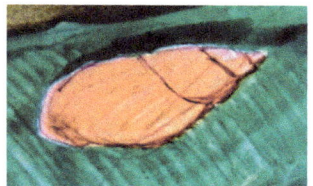

Partula exigua - the first victim

By the 1970s the three Professors had cause to worry that the *Partula* snails might be at risk. In the 1940s the Hawaiian Board of Agriculture and Forestry had decided to take action against the vast numbers of giant African land snails *Achatina fulica* (now more correctly known as *Lissachatina*). These giant African snails have a phenomenal reproductive rate, producing several hundred babies per year. That, combined with their voracious appetites makes them major pests in agricultural and garden areas. It is not surprising, therefore, that the population explosions of these familiar 'pet' snails have caused conspicuous damage to vegetable and fruit growing in the islands. The agriculturalists were particularly interested in the use of biological control: releasing animals that would eat or parasitize the snails.

Fig. 1. *Achatina fulica*, each of these adult snails is as long as a human hand

In 1948 two Kenyan carnivorous snails, *Edentulina affinis* and *Gonaxis kibwezensis*, were sent to Hawaii for research. While they were being studied the renowned expert on the giant African snail, Albert Mead, and Yoshio Kondo visited the islands of Micronesia to select one for a biological control field trial. They decided that it was a complex issue that needed a careful approach; they also noted that *Gonaxis* was probably too small to control *Achatina*, and they recommended consideration be given to larger predators. In 1950 the Insect Control Committee for Micronesia selected the small uninhabited island of Agiguan in the Marianas as the most suitable test site. Probably the most famous conchologist of his generation, the American R. Tucker Abbott was sent to East Africa to collect more snails. On 31st May 1950 some 400 *Gonaxis kibwezensis* were released onto Agiguan by Robert Owen. Owen returned to the island just over a year later along with George Peterson Jr. and J. Lockwood Chamberlin. Unfortunately, logistical problems limited them to only three hours on the island but this was enough for them to confirm that the snails were still there: they were breeding and had spread beyond the release point. What the effects on the giant African snails were they could not tell, nor did they have an opportunity to look for the native Agiguan snails. The following year Kondo, who had been managing the captive *Gonaxis* colony in Hawaii, was sent back for a more thorough survey and to release a further 100 *Gonaxis*. Kondo spent 17 days there, long enough to estimate that there were now around 21,750 *Gonaxis* on the island. Although he found them feeding on *Achatina* he concluded that *Gonaxis* was not controlling the giant African snails.

While these experiments were in progress on Agiguan, in Hawaii the Hawaiian Board of Agriculture sent its entomologist to East Africa, Australia and New Caledonia to look for other potential biological control agents. He came up with more *Edentulina* and *Gonaxis*, other carnivorous snails and beetles. The majority of these died in the laboratory in Hawaii. At the same time other people were looking for similar solutions to the *Achatina* problem. In 1953, in what proved to be a terrible indicator of problems to come, the Dutch agricultural consultant H.J. de Wilde de Ligny observed the impacts of a flatworm in New Guinea on *Achatina*: "It is believed that this snail is unable to cross the patches of jungle isolating the small farms from the town and from each other. This assumption is supported by the fact that a concentration of empty shells was often found in the outer regions of the forest surrounding the town. In a few cases naked snails [slugs or flatworms], probably parasitic, were observed on the shells of the giant snail. Experiments will be made to ascertain the carnivorous habits of the naked snail."

In a ghastly mistake Mead wrote:
"The large species of the voracious, tropical American *Euglandina* would seem to be worth serious consideration if biological control is to be investigated further. The eggs of *Euglandina* are

proportionately gigantic and the hatching individuals are therefore so large that by far the majority of the endemic molluscan fauna of the Pacific islands, consisting of small or minute species, would escape their ravages."

Thus he dismissed the fear being raised by some, Kondo included, that these predators could turn on the endemic Pacific snails.

Euglandina species had first been considered for introduction over 40 years before. In 1913 the Daily Mail newspaper had reported that the Mexican '*Glandina guttata*' (now *Euglandina vanuxemensis*) had been released in France. It was claimed that the predator was "conferring immense benefit on market gardeners." In reality members of the Société nationale d'Acclimatation were keeping the species in captivity and since 1910 had been evaluating whether or not it would survive if released into France; none had actually been released. Although some members bred the snails they eventually died out and when release into the plantations on Réunion was proposed in 1913 few were left. The last ones had gone to M. Vignal, an experienced breeder of many snails. After two years without breeding Vignal had only two left: "seeing that of the last two Glandines, one, the weaker, took little food, he decided to reunite them, hoping to obtain some eggs; the result was disastrous, the stronger devours its companion in captivity in a few minutes, preferring this food to that which it was served habitually and now she remains single, without probability of reproduction."

Even though the Insect Control Committee for Micronesia had planned their 1954 introductions to Agiguan carefully, while Kondo was studying the situation on the island, the Hawaiian Board of Agriculture and Forestry decided to skip ahead. They succumbed to pressure to deal with *Achatina* quickly and released *Gonaxis* on Oahu without waiting for the results of the Agiguan experiment. At the same time they released a predatory beetle. Ironically, arsenate bait put out for *Achatina* also attracted the *Gonaxis* and killed the entire release group. Undeterred, a further 500 were released in 1954, this time without the lethal bait. Mead inspected the site a few months later and found the snails well established. From there they were spread to Maui.

In 1954 Peterson visited Agiguan again and gave a more positive assessment than Kondo had done. It was now concluded that there should be a '*Gonaxis* Program' across the Pacific. So in 1955 over 5,000 *Gonaxis* were collected from Agiguan and distributed around American territories: the other Marianas islands (Saipan, Tinian and Rota), Hawaii, California (around San Diego) and the Caroline Islands (Truk [Chuk], Ponape [Pohnpei] and Palau), and to Australian administered New Britain (the largest island of the Bismarck Archipelago of Papua New Guinea). Although the snails adapted well to most of their new island homes, they had little impact and within two years interest in *Gonaxis* had waned.

All this while the Hawaiian Board of Agriculture and Forestry's entomologist was investigating different species. Krauss obtained *Euglandina* from Florida and released them onto Oahu island in November 1955, *Oleacina oleacea* from Cuba in 1956, *Gonaxis vulcani* and an unidentified *Gonaxis* from the Belgian Congo in 1956, *Gonaxis quadrilateralis* and more *Edentulina* from Kenya in 1957, and *Gulella wahlbergi* from South Africa in 1957. None of these was ever seen again, except for *Gonaxis quadrilateralis* and the *Euglandina*, which multiplied and spread rapidly

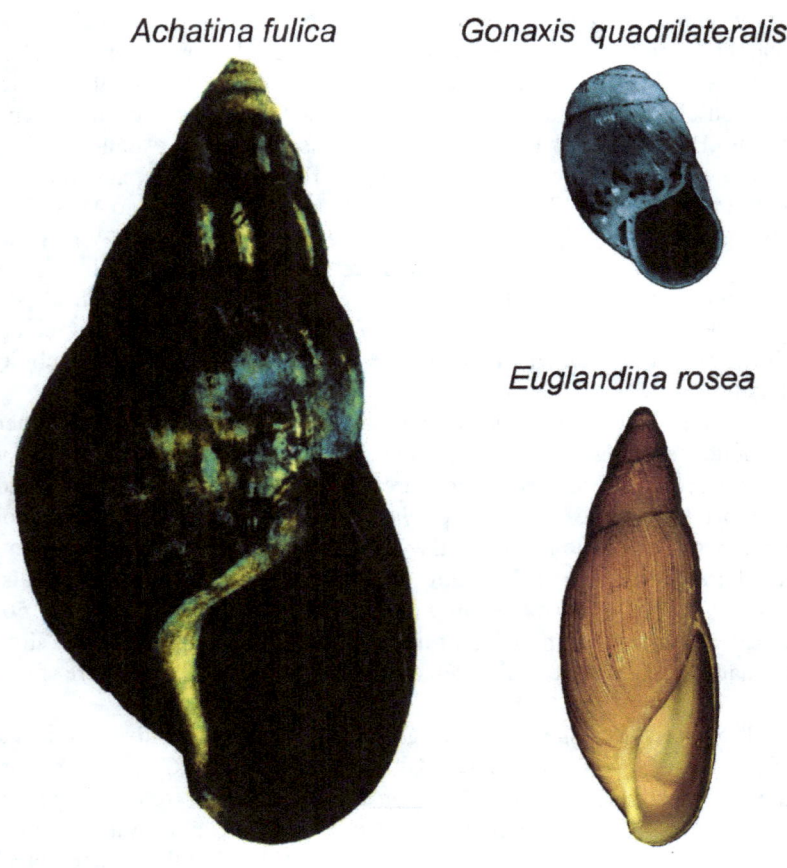

Fig. 2. The invasive snails, all shown to the same relative scale

Not everyone was delighted with these releases. As early as 1957 the botanist Raymond Fosberg was critical of the "almost hysterical program of importations of predators in an attempt to control the giant African snail". With great prescience he wrote "these will almost surely bring about destruction of many members of the extraordinary Hawaiian land snail fauna."

Of all the species released only *Euglandina* established itself and spread dramatically. In 1958 Yoshio Kondo was finding dead Hawaiian *Achatinella* tree snails littering the ground in the areas invaded by *Euglandina*. By 1966 populations of *Achatinella* were clearly in decline, several species had not been seen for several years, and have not been found since. The forests where tree snails were absent were now teeming with *Euglandina* and there seemed little doubt that the predator was the primary cause of their disappearance. In some places *Achatina* had become rare, but in others it remained abundant and was ignored by *Euglandina*. Despite the obvious damage that *Euglandina* was doing to the *Achatinella* snails, the predators were now on the move. In 1960 the University of California Agricultural Experiment Station had released 400 in Riverside County, California to control *Helix aspersa*. They did not do very well, although they were breeding there. The following year another 312 were released, along with 145 in Alameda County and 52 in Contra Costa County. Within a few years they had vanished, without having had any appreciable impact on *Helix*.

Despite the problems caused by the introduction of giant African snails to Hawaii, in 1967 they were imported to Tahiti apparently with the aim of farming them for food. They escaped and rapidly spread over the whole island and found their way to all the other Polynesian islands. The population explosion of these voracious snails inevitably caused pressure to control them. The unintended consequences of the introductions to Hawaii were pointed out to the French Polynesian Service de l'Economie Rurale, the body responsible for agriculture in the islands, by John Burch in 1970. However, *Euglandina* was introduced to Tahiti in 1974.

Euglandina was released at Papeari, Papara and Taravao in the south of the island. Concern over the probable impacts of *Euglandina* prompted Bryan Clarke to write to the Service de l'Economie Rurale, asking them not to introduce *Euglandina* to Moorea. They responded that they had not yet decided whether to do so or not. Despite this, on 16th March 1977 *Euglandina* was released on Moorea at Paopao.

On both Tahiti and Moorea the introductions succeeded in establishing *Euglandina*. From the introduction points they spread at a rate of more than a kilometre a year, a rapid expansion for a snail. When in 1982 Bryan Clarke and Jim Murray revisited the islands they were unable to find a single *Partula* in any of the sites occupied by *Euglandina* except for a few *exigua* in the east

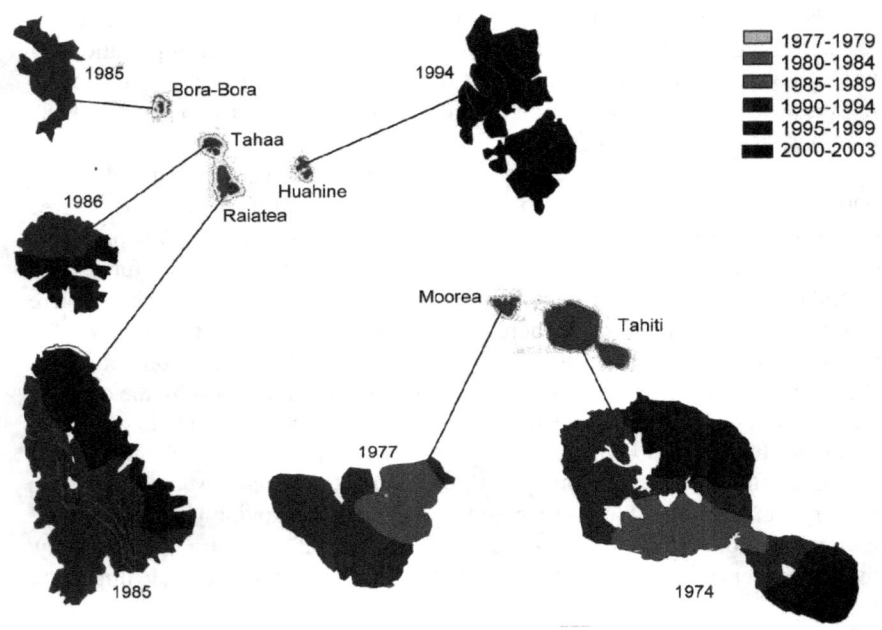

Fig. 3. Map of the spread of *Euglandina* across Polynesia, modified from maps prepared by S. Aberdeen.

of Moorea. By the sites occupied by *Euglandina* except for a few *exigua* in the east of Moorea. By this time a large part of Moorea was invaded and on Tahiti it had spread across all of the Temarua valley and Taravao plateau. When Bryan Clarke next collected on Moorea in November 1986 *Euglandina* had spread further and much of the island was devoid of *Partula*. He managed to obtain some *Partula* in Haapiti valley for his genetic studies but by then it was clear that the prospects for the snails on the island were very bleak.

        On both Tahiti and Moorea the introductions succeeded in establishing *Euglandina*. From the introduction points they spread at a rate of more than a kilometre a year, a rapid expansion for a snail. When in 1982 Bryan Clarke and Jim Murray revisited the islands they were unable to find a single *Partula* in any of the sites occupied by *Euglandina* except for a few *exigua* in the east of Moorea. By this time a large part of Moorea was invaded and on Tahiti it had spread across all of the Temarua valley and Taravao plateau. When Bryan Clarke next collected on Moorea in November 1986 *Euglandina* had spread further and much of the island was devoid of *Partula*. He managed to obtain some *Partula* in Haapiti valley for his genetic studies but by then it was clear that the prospects for the snails on the island were very bleak.

In 1987 the Captive Breeding Specialist Group of the International Union for the Conservation of Nature (IUCN) organised a meeting in London to plan the rescue of *Partula*. This meeting led to Jim and Bess Murray spending three weeks in June-July 1987 surveying the situation and rescuing whatever *Partula* remained. They searched 16 valleys, reaching high into the valleys in the hope that some populations might survive in the higher areas. They climbed all the way to the crests of the ridges of Paparoa, Faamaariri, Uufau and Maramu valleys. Shells were everywhere, but not a single living *Partula* remained; in each valley *Euglandina* had got there before them. Three valleys isolated by some form of barrier were also investigated: Tehaoa, Taapurau and Oio. Again *Euglandina* was well established and *Partula* already extinct. They concluded that the rescue attempt was too late; that no *Partula* survived on Moorea and that the last survivors had been those that Bryan Clarke found in Haapiti Valley seven months before.

The Murrays did not survey Tahiti in great detail but it was clear that *Euglandina* was already widespread there as well. They could only find *Partula* in the valleys of Tiarei and Mahaena (*otaheitana, affinis, hyalina* and *clara*). Huahine was still *Euglandina* free fortunately, and *varia* and *rosea* were still abundant. They did not find *arguta*, but that had always been scarce.

Although the only official introduction of *Euglandina* had been to Tahiti and Moorea the carnivore was spreading. From a relatively rapid crawl out from those introduction points, its range suddenly expanded. Private individuals hoping to remove giant African snails from their properties now picked it up and carried it around the islands. New populations appeared on Tahiti and by 1986 it had appeared on Bora-Bora and Raiatea. There was no information from Tahaa but it had probably arrived there as well. The last *Euglandina*-free island in the Society islands, Huahine, fell to the invasion in around 1992. At the same time it was found in the Marquesas islands of Hiva Oa and Nuku Hiva. Rurutu seems to have been invaded earlier than that, maybe as early as 1982.

The introduction of *Euglandina* was clearly a disaster for *Partula*, as it had been for *Achatinella* in Hawaii years earlier, but did the introductions achieve what they were supposed to? By the time *Euglandina* was released on Moorea *Achatina* was already in decline. After *Euglandina* appeared on the island the decline of *Achatina* was dramatic, even in the areas that *Euglandina* was yet to invade. *Achatina* was also disappearing on other islands, so, ironically, when it came to Huahine *Euglandina* was introduced to an island where *Achatina* had largely disappeared. Although the giant snails were no longer the plague they had been before, this cannot be attributed to the attentions of *Euglandina*. What happened in the islands is probably what tends to happen when an alien species is introduced: boom and bust. It may take some time for an alien species to adapt to the local environment and become established. Once that happens the population often explodes; it becomes invasive, spreads rapidly and as in the case of *Achatina*, may attain plague proportions. At this point it has escaped the limitations that ecology normally imposes on species numbers: local predators and parasites may not have adapted to its presence, diseases may be spreading only slowly and food is abundant. The population explosion cannot continue for ever; eventually food runs out, or at the highest population densities diseases spread dramatically. In the case of *Achatina* it seems to have been a common bacterial disease which caught up with them in the end. Even without *Euglandina*, *Achatina* populations were doomed to crash.

Euglandina itself has suffered the same fate: vast numbers of predators could be found throughout the islands in the early years of the invasion. By 2004 it had become rare on Tahiti and Moorea and was only found in six valleys on Tahiti between then and 2007. In general it remains fairly scarce today but every now and then there is a local population explosion which eats all the snails in its path before disappearing once more.

The only positive note in this sorry tale was that at least all the Moorean species were being kept in captivity. All, that is, with one exception: in 1986 *Partula exigua* became the first *Partula* to be lost for ever. The captive colonies of the other *Partula* species had been established in order to conduct Mendelian crosses and other genetic research, but now they provided the means of saving the species from extinction. At the time that the decline of *Partula* was first realised the Jersey Wildlife Preservation Trust had been given some *Partula* from the genetics colonies. Now more zoos were brought in to turn a research project into a professional conservation breeding programme. *Partula* conservation had begun, but was it too late?

Chapter 8. *Partula* conservation

Partula taeniata – the first for conservation breeding

Partula conservation started in 1981 when the first captive breeding colony was set up at the then Jersey Wildlife Preservation Trust. To start with, they were given some *taeniata* from the genetics study populations. These were the most adaptable of the Moorean species. The next year Jersey was given *mirabilis, suturalis* and *tohiveana* from the same source. They struggled to establish these species and by 1986 they had all died out.

The laboratory research populations had been kept since 1962 and keeping protocols had been devised that were effective for most species. They worked for *Partula* at least; for some reason *Samoana* had never survived in captivity for long. *Samoana* lacked interesting colours, patterns or chiral variations for genetic research and was very rare, so protocols were never devised for that genus. The methods were not complex and large numbers of snails could be housed in plastic boxes in a small room. In the early days most colonies were in small boxes, measuring 12x12x2.5 cm. More recently fewer, larger tanks have been found to be more practical. The diet was relatively simple, compared to many of the complicated diets zoos have to produce for their more picky animals. The snails did well on a diet based on porridge oats, grass or nettle powder, vitamins and added cuttlebone or chalk for calcium. The temperatures were not specifically designed to be the same as in the wild, but were what was available in the universities of Nottingham and Virginia. The zoo populations varied widely, from 19 to 27°C. In the wild the highest altitude sites occupied by *Partula* and *Samoana* may occasionally fall to 6°C at night, but species that occurred in lowland areas, such as *taeniata, hyalina* and *clara*, may need much higher temperatures. In Jersey the snails were kept in an office in the Reptile House where they were cared for by the Curator of Herpetology Quentin Bloxham and the newly arrived herpetology keeper Simon Tonge. This was Jersey's only invertebrate conservation project and the Reptile House the only place with suitable temperatures.

Even in the experienced university laboratories these conditions did not suit all species; no *Samoana* species or *exigua* ever survived long enough to be brought into the lab and *aurantia* quickly died out in captivity. For the species that could be kept there have been many attempts over the years to refine the methods. They have prospered when not overcrowded and the babies do well if separated to reduce

Fig. 1. The early days of *Partula* breeding - Jim Murray in the *Partula* breeding lab at the University of Virginia in the early 1970s. Photo: J. Murray.

overcrowding, although this is not always the case; excessive hygiene may be problematic. There is some evidence that juveniles follow adult slime trails and they may need to acquire their associated gut bacteria from adults. However, this is far from clear as at least one colony doing extremely well under conditions of strict hygeine

The most consistent finding though, has been that they do not like sudden change. The Nottingham colonies used to suffer sudden population declines whenever their carer, Vivien Frame, was absent. No matter how careful the laboratory technicians were, and how carefully they tried to copy her methods to the last detail, the snails would still die. For a while it was thought this might be down to something as obscure as a hand cream, but even controlling for this was right failed to stop the declines. To this day we don't know what the special factor was. What is certain is that it did make her rare holidays stressful, knowing that her snails would be pining to death in her absence! Since then, similar effects have been found with other *Partula* keepers; it seems that the more the keepers care about the snails, the more they thrive.

Looking after the *Partula* meant an ever-increasing work load, for at this time Bryan Clarke, Jim Murray and Michael Johnson were still adding to the range of species they had in their experimental populations. Having worked on Moorean species with great thoroughness for 20 years, by the 1980s they were starting to investigate Tahitian species as well. They brought the Tahitian *nodosa* into captivity in 1984 and added *hyalina* and *clara* 11 years later.

In the mid 1980s Jersey tried keeping *Partula* again and were joined by the Zoological Society of London's London Zoo and the Royal Zoological Society of Scotland's Edinburgh Zoo. As Jersey had found earlier, it was difficult to establish the snails at first. In 1986 the meeting at London Zoo to plan *Partula* conservation agreed that there was a need to expand the size and number of the captive colonies and to consider how they could be re-established in the wild. It seemed that there might be a possibility of establishing a protected enclosure on Moorea in association with the scientific research stations there, Le Centre de Recherches Insulaires de l'Observatoire de l'Environnement de Polynésie Française (CORIBO) and the University of Berkely's Richard B. Gump South Pacific Biological Research Station. When the Murrays searched Moorea fruitlessly for surviving *Partula* they also investigated the potential for this reintroduction site.

In 1986 the '*Partula* Propagation Group' was formed, coordinated by Paul Pearce-Kelly of London Zoo. London was well suited to this role, having had a dedicated invertebrate conservation department for 106 years. The Insect House where London's *Partula* were housed was an old building, dating back to 1913 and had been designed for display rather than for conservation. Even so, there was space for a dedicated *Partula* room. Tucked away in the back of the staff office in the basement a tiny room was lined with shelves and shelves of plastic boxes. In each box was a colony of *Partula*. Just as importantly it had staff who were able to concentrate more of their attention on the *Partula* than was possible at most other institutions. Although they had not been at the zoo for long before London first started working with *Partula*, the zoo had two highly motivated invertebrate keepers in the form of Paul Pearce-Kelly (who had been at the zoo since 1982) and Dave Clarke (from 1984). They were to be crucial to keeping the *Partula* programme going over the next quarter of a century.

Before long the snails kept by the new *Partula* Propagation Group were prospering and by 1990 the captive population had risen to 4,077 snails. The great majority were still in the Universities: Nottingham had the most snails (2,000 of seven species) and Virginia the most species (15, with 500 individuals, including the very last individual of *aurantia*). Among the zoos Perth Zoo in Western Australia was the leader with only 120 snails of five species. The eight zoos had half as many snails as the universities and, of these, Jersey still held the most of any zoo (700, although of just two species). London now had 200 (also two species). Much smaller numbers were in St. Louis, Edinburgh, Antwerp, Chester and Berlin.

As the snails continued to disappear from the wild through the late 1980s and into the early 1990s, the captive colonies became more and more important. In some cases there had now been several generations of captive animals. Today, the oldest surviving Society Islands lineages are *suturalis* (the ancestors of which were collected in 1980 and 1985) and *taeniata* (from 1981 and 1982); these have now been bred in captivity for at least six generations. Even older though is the population of *gibba* from Guam. These are all descended from a single snail collected by Jim Murray in 1972! Fortunately, this species finds self-fertilisation very easy and being founded by a single snail was not a problem. Numbers have fluctuated, but that single snail has 520 living descendants.

Not all species have been so resilient. After a slow start *affinis* reached around 400 individuals in 1997, before crashing to around 20 in 2004-2006, then back up to 653 by 2013. Similarly, *hyalina* was stable at around 100 at the start of 1992 and from there rose to a peak of 550 in 1997 before declining steadily to the present low numbers. *tohiveana* dropped from around 50 snails collected in 1982 to just four, before rising again to 900. *mirabilis* also declined to just six in 1990 but now stands at over 680. The most dramatic change of all was seen in the *nodosa* brought into captivity in 1984, which had dropped to just 8 adults in the early 1990s but has recovered, now numbering over 5,000 snails. Some populations have fluctuated repeatedly: one of the *taeniata* subspecies, *simulans*, stayed at low numbers for many years, rising above 50 in 1997 and then dropping to just 15 in 2000. Four years later they had risen to nearly 150 before dropping to seven in 2009. At the moment it is rising again. The same story is told for *faba* which was collected in good numbers in 1991 and 1992 and by 1995 there were 180, before falling into a decline, today this species is on the very edge of extinction.

With these fluctuations in captivity and the certain demise of the wild populations, collecting any survivors from the wild would serve to both bolster the captive populations and rescue any remaining animals before *Euglandina* ate them. In 1991 a conservation expedition, 'Operation *Partula*', sought to collect whatever survivors remained. The expedition was made up of Bryan Clarke and several people from the zoos keeping *Partula*. Two had long been involved with *Partula*: Dave Clarke from London Zoo and Edwin Blake from Edinburgh Zoo. Roger Klocek from the John G. Shedd Aquarium in Chicago was a new addition to the *Partula* world. They visited Moorea, Huahine, Raiatea and Bora Bora. As expected, no live *Partula* could be found on Moorea, although shells were found of *suturalis, taeniata* and *tohiveana*. It was already too late on Bora-Bora: only shells of *lutea* remained. Huahine still remained free of *Euglandina* in 1991 and *Partula* (*varia, rosea* and *arguta*) were still abundant. Necklace makers were still in business there and 'hei' production was a major occupation for the women of the island. On Raiatea *Euglandina* was spreading rapidly and it was clear that few species would last much longer. The first described of all *Partula, faba*, had always been the most widespread and common Raiatean *Partula* and this was still true, but they could find it in only a few valleys (Tevaitoa, Hotopuu, Haamoa and Tehehani). In the south they found little other than *Euglandina* but apparently necklace makers in Hotopuu valley had been collecting *Partula* relatively recently. They searched and although *Euglandina* were present they managed to find nine survivors of what had been the second most widespread species on the island: the pretty little white and pink *hebe bella*. On the west coast they found some *tristis* and in the same place they made an exceptionally lucky find, collecting some *turgida*. This was the rarest of all – back in the mid-1800s Garrett had described it as "excessively rare" and thought it was in the process of becoming extinct, Crampton had found a single animal and it had been so rare that Cooke had offered a reward for any on the Mangareva Expedition.

Fig. 2. Polynesian hei necklaces of *Partula* made from the Huahine species: top a mixture of yellow varia and purple *rosea* (along with pink *Littorina* periwinkles), bottom necklace composed of banded *rosea* only. Shells of *varia* (top) and *rosea* (bottom) are shown on the right.

Of all the species that might have been expected to be found this was the least likely; now at the last minute there was a chance to save it.

As Raiatea is a large island with some difficult terrain this brief expedition did not manage to cover the entire island. Doubtless, several species remained on the island but with *Euglandina* spreading everywhere this would not last long. There clearly was a desperate need for another rescue mission to the island which still had the greatest diversity of *Partula* anywhere on earth. This rescue on Raiatea was to be the focus of my own involvement in *Partula*.

Chapter 9. Raiatea rescue

Partula dentifera

The first time I encountered *Euglandina* (back in 1982) was during heavy rain in an overgrown garden. Crawling out of the undergrowth were two moderately large snails of a pinkish-orange colour with finely ridged slender shells immediately recognisable as *Euglandina*. The animals themselves are striking, with long grey bodies and long curling lips either side of their mouths looking just like long moustaches. These flick up and down as they use them to taste the ground for the trails of other snails. They are undeniably very attractive, elegant animals until their moustache lips pick up the trail of another snail. Then they switch into single-minded predators, moving with surprising speed along the trail until they find their prey.

I had been interested in *Euglandina* since 1981 when I first heard of their devastation of *Partula*. In 1988 I collected *Euglandina* from Seychelles (once again introduced to the main island to control giant African snails) to send to Ulster where the snail ecologist Tony Cook was studying their behaviour.

Fig. 1. *Euglandina* hunting

The transport to Ireland of these snails was arranged through Paul Pearce-Kelly. We remained in contact from then onwards and my ideas for work on *Euglandina* and *Partula* developed rapidly. In 1991 I started my PhD research on 'The ecology of the carnivorous snail *Euglandina rosea*'. I had followed the story of the extinctions in wake of the introductions of *Euglandina* in the Pacific islands and was struck by the observation that no such extinctions had been reported from the Indian Ocean where *Euglandina* also had a long history. I wanted to know why there was this difference. Had the predators failed to become fully established in the Indian Ocean; was there a difference between their prey in the different regions; or, worryingly, had extinctions in the Indian Ocean been overlooked?

In order to answer these questions I needed to study both *Euglandina* and its various prey species, and to investigate the situation in the wild. First I looked at the Indian Ocean islands, quickly finding three contrasting situations. In the Seychelles islands *Euglandina* had failed to spread, being trapped behind a barrier of poor habitat where prey were scarce. Beyond that zone there were plenty of snails, but these had evolved with indigenous carnivorous snails. They were fast breeders and had effective defences, being faster movers than most snails and wriggling to dislodge or deter predators. *Euglandina* had spread all over Mauritius, but there seemed to be little impact. There too the snails had their own defences. Perhaps more significantly, what little remained of the Mauritian forests were full of invasive species, from just about every plant ever introduced to a tropical island, to a great range of snails and insects, to pigs, deer and monkeys. There had been many snail extinctions there, but all could be blamed on the loss of good habitat, rather than on *Euglandina*. On Réunion island natural defences seemed to protect the native snails. In addition *Euglandina* had not covered all of Réunion, most notably it seemed slow to invade the upper parts of this mountainous island.

In the Pacific Ocean there were two places that reports indicated were of interest: New Caledonia, where no extinctions had been reported, and the Society Islands. New Caledonia turned out to be like Seychelles: *Euglandina* was stuck in bad habitat, at least for the time being. Studying *Euglandina* in the Society Islands was an opportunity to test a theory I had developed. By this time my laboratory studies had shown that *Euglandina* was not able to tolerate cooler temperatures, which seemed to explain why it had failed to cover all of high Réunion. I now wanted to see if the same might be true in the Pacific. If this was the case there might be a possibility of some *Partula* surviving at high altitudes, if they could tolerate the cool temperatures. Raiatea was the ideal island to visit. There I could test my theory, study an ongoing invasion and rescue whatever was left before *Euglandina* covered the whole island. Two months before starting my Pacific fieldwork I telephoned Bryan Clarke. During the course of a very interesting conversation he told me that one species of *Partula* had just been seen on Tahiti. It seemed that there was a population of *otaheitana* on the mountain overlooking the airport. It sounded to me like the first evidence of a *Euglandina* altitude ceiling from Polynesia, so I was eager to investigate further.

I arrived in Tahiti in the night of 19th August 1992, and stayed in a hotel on the outskirts of the town of Papaetee. As Raiatea was to be the focus of my work I had only a few days on Tahiti. In exploring Papaetee I found that the market included many curio stalls selling shells, mostly in the traditional form of necklaces. Not surprisingly none of these were *Partula*, and one stall was even selling varnished *Achatina*, testament to how the fauna had changed over the previous decade.

My impression from Papaetee was that Tahiti was in a much worse state than I had expected. The dramatic mountain behind the town seemed to be almost completely tree-less. Everything near the town seemed to be introduced, from the plants to the stray dogs and the common red-whiskered bulbuls, mynahs, ground doves and feral pigeons. On the 21st field-work started, with a walk up the road to the viewpoint at Fare Rau-Ape. The vegetation was open secondary growth and scrubby grass on the mountain slopes. There were trees of *Hibiscus tiliaceus* on which *Partula* had previously been reported to like feeding on, and the invasive *Miconia calvescens*. *Miconia* is one of the most striking features of these islands, with enormous leaves reaching almost a metre in length, appearing (but not feeling) slightly velvety, dark green above and a deep purple underneath. Some of these shrubs must have been 10 m tall. Small, light grey rats were everywhere, scurrying up the banks or just squashed on the road. There were some native reptiles around: a large brown gecko hiding under the bark of a tree and a species of very wary skink that was initially difficult to spot despite their golden striped flanks and bright blue tail. There were more birds: grey-backed white-eyes, Society fruit doves, red-browed and common waxbills, and feral chickens! *Euglandina* shells were common and I found one alive, but other snails were very scarce. I found around half a dozen small leaf-litter species but no trace of any *Partula*.

The 24th was to be my big snailing expedition on Tahiti, the search for the surviving high-altitude *otaheitana*. In contrast to previous days of rain it was a brilliant cloudless morning, with Moorea standing proud with its savage rugged beauty, a big shape of spikes sitting heavily across the near horizon. There was a long walk along the motorway to the road leading to the view-point at 1400m which I hoped would be above the altitude ceiling for *Euglandina*. The road carried on, climbing up though suburbs behind a surprising number of garbage trucks. I spent the day searching for the right road, but found nothing at all beyond suburbs and garbage ending up at the municipal rubbish tip. Eventually I give up the excursion as a wasted effort: finding survivors on Tahiti would need a special search with proper transport. It was frustrating but as Raiatea was the main focus of my planned research, failing to find *otaheitana* was not of great importance.

The next day I moved to Raiatea. PhD research is always done on a tighter budget than is sensible and the ferry was considerably cheaper than flying to Raiatea. As a ferry the 'Teporo IV' was not what I expected, being a moderate sized cargo ferry with some limited deck space for passengers, a handful of cabins somewhere and an indoor area bordered by a row of fixed plastic seats. This space was covered with mats as the first people on board seized what was presumably the dry sleeping space. Out on deck were two rows of hard seats. I opted for the open deck rather than the already stuffy indoor space, even though I suspected the deck would be wet.

The cargo of sacks of copra had been unloaded before I arrived. The warm, sticky and slightly burnt smell of copra was very strong, somewhere between thickly appetising and nauseatingly cloying, becoming more and more unpleasant as time went on and we sat in the harbour. The return cargo seemed to be predominantly toilet rolls. The ferry filled up with many small children, then a group of young drunks who were having trouble standing. Once the ship sailed they settled down on deck to carry on drinking and started strumming guitars and a banjo. Despite their condition they played and sang reasonably well. So we sailed, with Moorea to our left, the sun setting slowly and Venus growing brighter, to the sound of Polynesian and Franco-Polynesian songs.

During the night the crew ordered us inside; it seemed they were expecting rough seas. We were getting a little splashed and disturbingly the toilet rolls in the hold could be heard listing from side to side below us. Inside there were no stars to watch, the air was stuffy and high from the people already there. The chairs were hard but it seemed I was the only one not to fall asleep instantly.

We stopped briefly at Huahine in the early hours, then as the dawn glow started across the sky Raiatea and Tahaa emerged from the cold darkness. My first thought on seeing Raiatea was 'Oh my God it's steep.' The port was even more miniscule than the photos I'd seen and the town of Uturoa not visible. I was met at the port and driven to the Pension Greenhill. Driving down the coast I found the views of the hillside vegetation not at all as I expected. Coconut plantations extended too far back for my liking, meaning that all useful habitat was a long walk back and up the mountains. Giant invasive *Paraserianthes* trees predominated and the lowland areas were smothered by *Merremia* creepers. This was not promising for finding native snails but, on the other hand, the terrain did not look too bad. At Faaroa bay the Pension Greenhill stood above the road, with its spacious garden and splendid view. There I was taken in hand by the owner, Marie-Isabelle an exuberant and eccentric Indochina born Frenchwoman. She immediately plied me with a great breakfast. She was somewhat concerned about the length of my stay as there was nothing to do, but when I explained my purpose she was delighted – I was a friend of her 'snail men'! Within minutes she was trying to get hold of the mayor, and if he had been available at that hour (it was barely 7 am) he would have been ordered to organise everything for me. Fortunately for him, he was away from the island for the week. She showed me the *Partula* shells the 'snail men' had given her, laughing with the memory of them treating them 'like they were platinum', and found me a live *Euglandina* from the garden. All before I had finished my breakfast. This seemed very promising for my research; even if all of the *Partula* had gone already *Euglandina* were common.

I settled into my charmingly tatty room at the back of the garden, nestling into the jungle of the hillside, then spent a couple of hours looking for shells along the nearby bank. In just five metres I collected a few species, three live *Euglandina*, 12 perfect *Euglandina* shells and 40 recent eggs. "they are amazingly abundant; I don't know what I'll do if I find an advancing front!" I wrote at the time.

The following day my breakfast was accompanied by a couple of upside down tin mugs which concealed two *Euglandina*. I walked from the Pension hoping to climb the hill on the far side of the bay. It took me an hour to reach the river that bisects the bay, walking fast, only stopping to photograph the splendid mountains in the centre of the island. The mountains are very dramatic, stunning in their alarming severity. They don't inspire climbing, just wonder. The river area of the bay and the valley were a broad tangled mass of *Merremia* trailing across the flat and

Fig. 2. Faaroa bay from Pension Greenhill

festooning the tress. By the side of the road I made an exciting find – my first *Partula*. Below an area of slope cleared for manioc there were several *Partula* shells, all empty and none fresh. They were bleached white although some looked to only a year or so old. This offered hope that I might find some alive on the ridge.

Further on I found a way up the hillside into the *Hibiscus*. I found the vegetation to be dense, with many dead stems to crash through; slipping, falling, hauling and tearing at branches. There I found a few live *Euglandina*, but no *Partula*. To reach the ridge necessitated a charge up a steep slope through bracken that blanketed the ground under pines. There was more crashing through ferns and then more interesting-looking *Hibiscus* tangles. As I headed towards the crest of the hill I scrambled in and out of the pines and *Hibiscus* before falling into the depression formed by an overgrown forestry road. This road was covered in *Euglandina* shells, the numbers were unbelievable – thousands of them, perhaps millions without overstatement. I picked up a few more live. In the *Hibiscus* tangles there was little else; many *Achatina* shells but no trace of *Partula* at all.

Back at the road I returned to the location where I had found the *Partula* shells and scrambled up the bank there. This time it was *Hibiscus, Paraserianthes*, coconut and *Freycinetia*. Here there were plenty of shells of *Partula, Achatina* and *Euglandina*, but not one alive. Clearly *Partula* were already extinct there. It would have been wonderful to find live *Partula* on the first attempt, but it was good to find the shells. As I wrote that day "I hope, but do not really expect to find living survivors."

I was given a lift back to the Pension, detouring around the south. I took the opportunity to spot potential field sites. I could see from the truck that a large area on the slopes of Faaroa bay was dominated by bamboo and pointless for surveying. The mountains looked fascinating but off-putting. The reason no-one had surveyed the ridges for *Partula* was that no-one had been quite mad enough. I thought that with a helicopter it could be possible, otherwise any ascent would be insanity. Fortunately not everyone took my view, and some years later some of those areas were explored, with interesting results. The truck driver was very talkative and generous in his offers, he wanted to know all about me, snails both amused and puzzled him, did I know his friend from England?, was La Reine Elisabeth from England?, were there many people there?, did they fish for supper? He told me all about the joys of pig hunting, offered to take me fishing or to help snailing, offered to cook a traditional Polynesian meal for me and told me to tell any Jehovah's Witnesses that I was a friend of one of their Polynesian brothers.

The following day was a diversion, with a trip organised for the Pension guests to the little coral islet of Motu Irirui on the edge of the island's fringing reef. The trip across to the islet was in a little lighter; there was just enough swell to make it a wet crossing. The island is a tiny flat scrap of sand with a couple of hundred very tall coconut trees. There was no undergrowth and only the odd patch of grass. The ground was covered in very large crab holes, made by the enormous land crabs that run all over the roads on Raiatea. As the Forsters had found on the same islet in 1774 there were no shells of interest there.

I surveyed several places in the vicinity of Faaroa Bay, in all finding live *Euglandina* and shells of *Partula*. Most memorable of the sites in those areas was up the river that feeds into Faaroa Bay, through open scrubby areas, plantations of grapefruit trees, taro and banana. All were quiet, with no-one around in the rain. After an hour I left the track and followed the river upwards through the bamboo.

In the bamboo forest there were a few *Euglandina* foraging. That was a good start for my purposes, although not hopeful for *Partula*. Although I usually hate bamboo for its irritant hairs and dangerous shards, I enjoyed myself crashing about in the rain. The damp seemed to soften the edges of everything and the spiky bamboo felt less threatening than it usually does. In one area it gave way to *Inocarpus fragifer* trees growing around the river. These 'mape' or Tahitian chestnut trees were splendid and the atmosphere in that patch raised my already high appreciation of Raiatea. The trees were tall and fissured, moss strewn. The bases are supported by wonderful buttress roots, high and narrow, pleated and meandering in swirls over the rocks. They look like some sort of wayward over-elaborate pastry, like something out of a surrealist nightmare or Tolkien's imagination.

Ferns were profuse in the damp, including birds nest and tree ferns. There were some *Procis* and *Freycinetia*, patches of delightful variation before the reversion to bamboo, now a green hell of close fitting stems. Brittle dead bamboo stems cross the gaps, pit tracks and river. The Lonely Planet guide I had with me accurately described the end of the river as 'heart of darkness' jungle, however, they were describing the end of the navigable river much lower down, not this area. Some time later I could see light through the tangle ahead and only slightly above me. As the river course became steeper approaching the ridge, the vegetation changed from pure bamboo to bamboo and *Hibiscus*, then to hygrophilic ridge vegetation. This was carpeted in moss, dense with ferns and undergrowth, with *Pandanus* and *Freycinetia*, tall trees with epiphytes and water dripping everywhere. Returning from there I found five shells of a fourth *Partula* species.

A visit to the small, scruffy town of Uturoa gave me the opportunity to survey the snails between there and Faaroa Bay. For a snail collector no roadside walk in any new place is wasted, as the gutters provide an excellent collection of anything living nearby. It might not have been expected that this would be particularly useful but the snails in the gutters caused me a lot of consternation. There were few shells in Uturoa, only *Achatina*, but these were recent. Although I found it hard to accept I had become used to the idea that in Polynesia live or recent *Achatina* were virtually non-existent where *Euglandina* were common. The abundance of *Achatina* suggested

Fig. 3. Mape tree

Euglandina might be absent. This went against all my expectations; all my preconceptions had been based on the assumption that Uturoa was the introduction site. If this were to be wrong it would mean that I was searching for *Partula* survivors in the wrong places and my whole survey plan would need to be turned inside out.

Walking south I continued to fail to find *Euglandina*. I found some shells but all were *Achatina* and *Bradybaena*, the latter a species easily eaten by *Euglandina*. I was getting increasingly worried but I could not believe that I could have got it so wrong. Eventually I found a *Euglandina* at Avera where the broad northern plateau shrinks and the mountains fall directly into the sea, and that seemed to make sense. The origin could still have been Uturoa and the absence of shells a reflection of the dryness and openness of the plateau up as far as Avera.

Further surveying was organised in collaboration with Jean-Pierre Malet from the Economie Rurale, whom London Zoo had recommended to me as a most helpful contact. First we explored the Opoa valley, accompanied by Emile Brotherson. As usual, this valley started off with *Hibiscus* and coconuts, becoming increasingly diverse as we moved up and inland. There were many *Euglandina* about and lots of old *Partula* shells, but no evidence of any survivors. These *Partula* were recognisable, being the pretty little *hebe bella* that the previous year's expedition had rescued from the same area. In the pouring rain we struggled through the vegetation and then started straight up the ridge, through thick bracken and tree ferns; very steep and slippery in the downpour.

The next day Jean-Pierre turned up with a different assistant, an enormous Polynesian man with the most dramatic drooping Mexican bandit moustache. Again it was good, wet snail weather. We drove round the south end of the island stopping at various points. The first place we stopped we found *Euglandina* and little else, not even any old *Partula* shells. The second stop took us into the heart of the valley opposite Faaroa valley. This was relatively open, and trampled clear by cows; again there was nothing of interest.

The third stop was in a tangle of *Hibiscus* and coconuts. It was not very inspiring but almost immediately I found that the ground was littered with surprisingly recent *Partula* shells. There were two species, one very different from any I had found before: little pale brown shells that I later identified as *garrettii*, the species named after Garrett by Pease in 1864. This had been restricted to just a few valleys on the island's west side. More abundant than the little *garrettii* were the large yellow and brown *faba*. This original *Partula* had been all over Raiatea and I had found shells on the east side, but all old, bleached or broken. Here most were in perfect condition; the population seemed to have been wiped out very recently. In fact, some shells still had smears of dried mucus on them, evidence of that they had been killed only days before. The 'bandit' even managed to find one living *faba*. This was a great morale boost, but we could find no more than this one animal. It was not much of a rescue but at least this last survivor could be added to the *faba* colony that had been breeding at London Zoo since the year before.

We completed the circuit of the island, but did not stop again until we were just north of Greenhill. It was an unlikely spot, being just a few hundred metres north of Avera where I had found *Euglandina* on the roadside. We had passed some other tracks heading inland but these all seemed to go into suburbs or dry hills. This last site for the day was at least vegetated. We drove up a very rutted track and stopped by a patch of pollarded *Gliricidia*. These little trees had been cut low as living supports for vanilla plants in a very small plantation beside some tall trees. The *Gliricidia* were covered in live *Achatina* and, more excitingly, *Bradybaena*. The snail itself was not of interest, being one of the world's most widely introduced snails, recorded as a pest of coffee plantations as early as the mid 1800s. Its presence was, however, of interest as the small *Bradybaena* was one of the favourite prey of *Euglandina* and I took this to be an indication of the absence of the predator: I had not managed to find a live *Bradybaena* anywhere that *Euglandina* had reached. To my delight, just a few steps from the track there were living *Partula*. What was more, it was not just one like the *faba* we had found earlier, but vast numbers. Here was a relict of the astounding abundance that had been found by Crampton and Garrett, and that Bryan Clarke remembered from Moorea.

Fig. 4. The dead *Partula* left behind by *Euglandina*. The majority are *faba*, with a few *garretti*.

Even though the habitat in this vanilla plantation was completely unnatural, the *Partula* were amazingly common. At the time I had no means of identifying what I came across, other than the distinctive species in London Zoo. Again *faba* was recognisable, and along with it a second species which lived under leaves on the ground. When I compared the terrestrial snails to animals in London Zoo later it proved to be the same as those collected in the same valley the year before and labelled *dentifera*. There were a few other snail species, and not a single *Euglandina*.

We stopped collecting when I had run out of space in my containers; these 50 *faba* and 13 *dentifera* were a far cry from the complete extinction I had been expecting. I wrote that day "It seems that I did get the pattern of spread completely backwards after all." It took a long time to sort out the collections, to transfer the living animals to new boxes and to see if they would take the artificial *Partula* food. There was no need to worry about their feeding, for they took to the new diet within seconds of being put in the plastic boxes; they seemed very adaptable.

Later we surveyed the valleys between the Pension and the *Partula* site of the previous day. These four valleys were mostly mixed secondary vegetation, without exception the tracks we took were all heavily overgrown and disused and Jean-Pierre's four-wheel drive was essential. Some of the places seemed unsafe even

Fig. 5. The last wild *Partula faba*

or this vehicle, but we struggled up the tracks none-the-less. However, the valleys had little of any interest: *Euglandina* were not common but they were present in all the valleys. Even so I did find a single *faba* on a guava tree.

A morning's collecting work was now generating more than a full afternoon's sorting and so I set aside two days to catch up with the material and to make sure the live snails were cared for. They all adapted to the dried food and as soon as the lids were back on the boxes they were rushing up to the food. Within a couple of hours all I had provided them would be gone.

The final survey was an expedition to the high plateau of Mt. Temehani, this time with Jean-Pierre and the 'bandit', plus 20 field workers going to attack the *Miconia* on the upper slopes of the mountain. As we rattled, bumped and lurched our way up the track I peered at the vegetation trying to spot tree snails, but all I could see were old *Euglandina* shells next to the road. Above the pine plantations we stopped and the machete sharpening started. That took quite a while until all were satisfied. We set off at a great pace until we came to a clearing overlooking the valleys below and just below the plateau, here the field workers all sat down. After a short while waiting for some activity Jean-Pierre, the 'bandit' and I left them there

Fig. 6. The last wild *Partula dentifera*.

and made one of our now familiar rash scrambles up precipitous slopes, leaving the field workers to descend the valley below and seek and destroy *Miconia*. At this time *Miconia* had only recently become established on the island and was restricted to only 2% of the island. In the next four years Jean-Pierre oversaw the removal of 600 mature trees and 645,000 plants, saplings and seedlings. It is still present on Raiatea but has been held from dominating the island.

We searched a small area of mixed vegetation hiding an unpleasantly steep slope. There were a few shells but little life and certainly no live *Partula*. All of a sudden I found myself clinging to a bare vertical rock somewhere above a valley hidden in cloud below me, wondering how the others had got up and whether the cause of conservation really required that I follow them. Follow them I did, with the view that it was best not to think about how we were going to get down. Of course, on the other side of the rock there was a perfectly good path; but what's the point of using the easy route?

From there we were onto the plateau. This folded hilly area of bracken fern, small bushes and heather looked like a rather hilly British moor in the low misty cloud. The impression held as long as the glimpses of Tahaa and the

Fig. 7. *Miconia clavescens* on Tahiti in 2005. It has not reached this level of dominance on Raiatea due to the vigilance and efforts of Jean-Yves Meyer and Jean-Pierre Malet and their staff (photo: J.-Y. Meyer, Délégation à la Recherece de la Polynésie française, Tahiti).

Fig. 8. Temehani plateau. Top: habitat of the plateau, with a thicket of *Pandanus* in the foreground. Bottom: the Temehani *Partula – labrusca*

tropical lagoon were avoided and the valleys ignored. The stream valleys held tangles of tree ferns and *Pandanus*. We searched one of these tangles without any luck although I did find a rather fine orange stick insect.

More traipsing over heather-like shrubs led to an area where Jean-Pierre had seen *Partula* the year before. This particular thicket was sheltered from the strong winds that swept across the plateau and it was very damp, very epiphytically mossy. There were still a good number of snails: six of the slender, sticky *Samoana attenuata*, another dozen or so *faba* and six of a species new to me. These round, dark *Partula* later turned out to be *labrusca*, a high-altitude species restricted to Temehani. It was a most successful end to the field-work.

With my 88 live snails I returned to Tahiti, passing the savagely sharp peaks of Moorea. Tahiti was cloudless again but the comparison was not favourable: a dry cone of brown scrubby vegetation, fissured by gully-like valleys and fringed by roads and houses. The idyllic Tahiti that so impressed Cook, Banks, Solander and the Forsters has certainly changed over the past 250 years. I felt though that I had found a trace of it in Raiatea.

Chapter 10. The nature of the beast

Partula producta

Although rescuing the Raiatean *Partula* was an important part of my fieldwork, the main purpose of my research was to investigate *Euglandina*. My aim with the aim of understanding it and, hopefully, identifying something that could be used to save *Partula*.

My research on *Euglandina* confirmed what we already knew: *Euglandina* is a phenomenal predator of tree-snails. It is fast, for a snail, easily able to outpace almost any other snail and certainly had no difficulty catching a laid-back *Partula*. With greatly elongated lips, drawn out into curling moustaches, *Euglandina* is able to detect prey slime trails and follow them to its victim. There its preference is to swallow the prey whole and so obtain the calcium it needs for its shell. If the prey is too large for this to be easy, the *Euglandina* rips off pieces of the snail with its steak-knife-like teeth and will

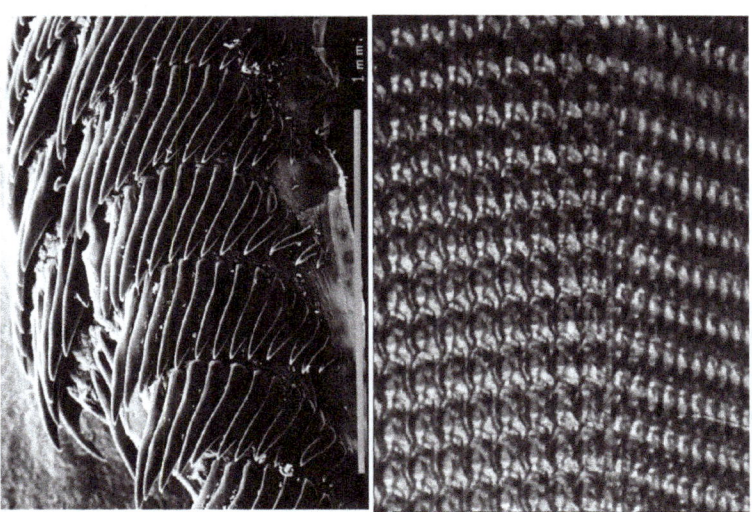

Fig. 1. The teeth of the *Euglandina* radula (left), compared to the typical grazing teeth of *Partula mooreana* (right)

then spend up to 30 minutes trying to cram the shell into its mouth. Most *Partula* are a perfect mouthful, the larger ones, like *faba*, are too large for swallowing. As I had found on Raiatea, where *Euglandina* had passed through, the ground was littered with *faba* shells.

When it comes to feeding on ground snails *Euglandina* is not so impressive. Although any snails living on the ground, including *Partula*, are easy prey *Euglandina* tends to be inefficient at hunting small snails in leaf-litter. Whereas it could find and attack almost all *Partula* on trees, it found no more than around a third of snails hiding in the litter. In deep leaf-litter it was even less efficient. As *Partula* only ever lived on the surface of the litter they were all vulnerable, but most leaf-litter inhabiting species live under the leaves and so receive some protection. Whilst this may be good for those snail species it raises a futher problem for *Partula* for there is little prospect of *Euglandina* populations dying out. Even after the large prey have all been eaten to extinction some predators will persist, feeding on the small litter snails, but never exterminating them. It had been hoped that *Euglandina* might disappear after the loss of *Partula* but it was not to be.

In this research I was hoping to find something that we could use to stop *Euglandina* or at least to give *Partula* a chance, but there was nothing. Anything that would be bad for *Euglandina* would be just as bad for *Partula*. These snails had no defences; there seemed to be nothing going for them and it was hard not to feel that they were the giant pandas of the snail world – cute but doomed, at least in the wild. There were only three points that offered any hope at all. Many snails seemed susceptible to a disease that was thought to be caused by a common bacterium. *Achatina* and *Euglandina* in particular could often be found with symptoms of discoloured skin. Severely infected animals would stop

Fig. 2. Skin lesions symptomatic of a disease in *Achatina*. Similar symptoms are seen on *Euglandina* and sometimes on other snail species.

feeding and an autopsy would reveal that their digestive systems seemd to have collapsed. Tree snails did not seem to be affected. Maybe disease outbreaks explained why *Achatina* populations sometimes collapsed for no obvious reason and maybe the same would happen to *Euglandina*. Modelling of the populations showed that it was unlikely that disease would wipe out *Euglandina* completely but it might reduce densities to such an extent that *Partula* could return one day. The models did not support the last part, but it was a glimmer of hope.

Secondly, *Euglandina* needed quite a lot of food. If one went several days without feeding it became lethargic, conserving its resources. In a vicious cycle, the less it moved the less food it could find, and so it would move even less. If there was a wide prey-free area, that might act as a barrier to *Euglandina*. The problem was that it had to be a very wide barrier without a single small snail in it. At the time there was talk of clearing the forest in one of Moorea's valleys for a golf course. While the destruction of the forest was to be deplored, this might have create such a mixture of trees and open areas. In the end the golf course was in an area totally unsuitable for *Partula*, but briefly it seemed to offer some hope.

The last point for possible optimism was that *Euglandina* did not like the cold. Above 18°C was its ideal; if temperatures dropped below 14°C it stopped being active. This explained why it was slow to invade high-altitude areas and supported my idea that the very highest parts of Tahiti and Raiatea might remain *Euglandina* free. I estimated that above 800 metres above sea level might possibly be its limit, but 1000 metres was more likely. Unfortunately, as far as we knew *Partula* did not like it up there either; certainly we were keeping the captive animals above 17°C on that assumption.

More recently we have found that the 'altitude ceiling' can be breached. *Euglandina* is common below around 900-1000m (varying on different islands), but small populations can occur to about 1200m. So far these have been temporary populations and once a *Euglandina* was found just below 1400m.

Euglandina is the classic example of how not to control invasive species. By trying to control *Achatina* by introducing *Euglandina* one problem was simply swapped for another. This catastrophic failure has not been the end of biological control in Polynesia however. Much more recently biological control has been used against the dramatic purple-leaved invader *Miconia clavescens*. This time a much more specific control agent was used, to better effect.

In the space of several decades *Miconia*, the 'bush currant', 'velvet tree' or 'purple plague', spread across much of Tahiti. This small (12m) tree is naturally restricted to the rainforest of Central and South America. There it is found under the canopy and is unable to dominate any areas except for temporary gaps in the forest. Its large purple leaves attracted botanical interest and it was deliberately planted in botanical gardens, first in Sri Lanka in 1888. Seeds from Sri Lankan plants were planted on Tahiti in the Motu Ovini private botanic garden in the southwest of the island in 1937. It was also planted at the Agricultural Research Station on the island a little while later. In 1961 it was planted in Hawaii and has since been introduced to New Caledonia, Queensland, Jamaica and Grenada. In 1996 it appeared in the French Polynesian Marquesas islands, probably as a result of seeds being carried by road building equipment from Tahiti. It may have spread from Tahiti to Moorea and Raiatea as seeds on vehicles, shoes or in pots.

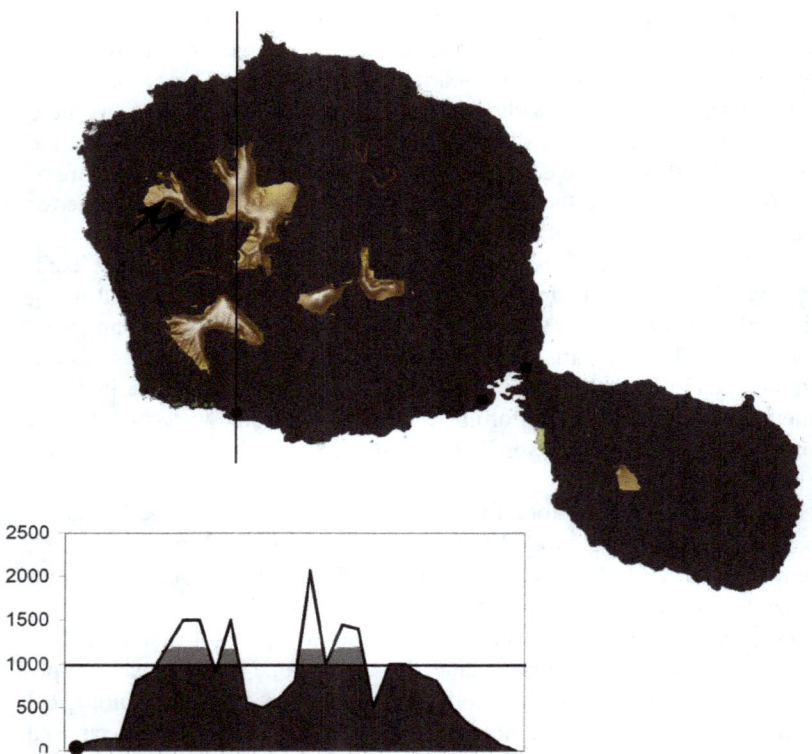

Fig. 3. Distribution of *Euglandina* on Tahiti in relation to altitude showing the apparent 'altitude ceiling'. Map shows range of *Euglandina*. Introduction sites marked as black points. Arrows show occasional records of *Euglandina* above the 'altitude ceiling'. Lower figure shows altitude profile (south-north) along the line on the map. Dark shading – *Euglandina* invaded areas. Light shading – *Euglandina* invaded areas above the predicted level of the 'altitude ceiling'.

Outside of its natural range it is a very aggressive invader, growing quickly (some 1.5 m a year), reaching maturity early and being able to reproduce through self-pollination, producing berries that are attractive to birds and contain enormous numbers of seeds; all features of successful weeds. The plant has a great impact on the environment due to their large size and rapid growth. By the early 1970s *Miconia* had spread widely over Tahiti and had colonised Moorea. It was not reported to be invading Raiatea until 1988, although it had been planted there in 1955 and had

Fig. 4. The dramatic impact of *Miconia calvescens*: Jean-Yves Meyer in a forest of *Miconia* at Pic Vert on Tahiti in 2004 (photo provided courtesy of J.-Y. Meyer, Délégation à la Recherece de la Polynésie française, Tahiti).

started to escape by 1970. In the 1990s it also appeared on Tahaa.

In the space of 50 years it spread over 70% of Tahiti and 25% of Moorea. In many areas it forms dense stands where its closed canopy shades out all other species. These can be extremely dense, with up to 6 plants per metre. The invasion of *Miconia* is considered to threaten almost half of the plant species restricted to Tahiti. The dramatic change it has caused will also be having an impact on the animals living within the affected habitats, including the *Partula* snails. As the invasion started in the south of Tahiti the *Partula* of that part of the island would have been the ones most affected. Most vulnerable of all would have been *producta*, a single species restricted to a single valley in the south of the island and there facing the double threats of habitat change and the *Euglandina* invasion.

There problem of *Miconia* on Tahiti was recognised by the 1970s but research on the species did not begin until the start of the *Miconia* Research Program in 1988. Jean-Yves Meyer's research on *Miconia* concluded that it could not be effectively controlled on Tahiti due to the scale of the problem. At this time Raiatea was not so badly affected: it was restricted to three valleys and Miconia patches were largely surrounded by pine plantations. Here physical removal seemed still to be practical. In 1992 25,000 plants were removed by hand (as I had seen for myself) and 77,000 the following year. Continued intensive weeding by a total of 3,500 people (now totalling over 2.5 million plants removed) has largely restricted the problem, but new plants continue to be found and it is clear that even this level of effort is not enough to stem the invasion. On Tahiti and Moorea, by the 1990s the scale of the problem was already far too great for this form of control.

Control of *Miconia* has proved extremely difficult due to the extremely steep terrain of the islands and the scale of the problem. Weeding and chemical treatment have been ineffective and other approaches were investigated. In order to deal with the invasion of Hawaii, the Hawaii Department of Agriculture searched for potential biological control agents from *Miconia*'s natural range in 1997. Various possible biological control agents were investigated: fungi, nematode worms, beetles, wasps, butterflies and moths. The most promising species to be identified was a Brazilian fungus *Colletotrichum gloeosporioides* f. sp. *miconiae*. The fungus species *Colleotrichum gloeosporioides* is not a host-specific pathogen; it infects a wide variety of plants, with many different specialised forms. However, laboratory tests showed that this particular strain of the fungus was highly specialist, pathogenic to *Miconia* and not affecting any other plant tested. This indicated that it would only affect *Miconia* and so should be safe to release on islands that lacked any native *Miconia* species.

The fungus causes necrotic spots on the leaves, which turn into lesions and holes. It also causes dieback from infected stems and kills *Miconia* seeds and seedlings, with up to 74% mortality. Jean-Yves Meyer proposed that it should be

Fig. 5. Map of the main populations of *Miconia calvescens* in French Polynesia. *Miconia* range shaded on relief maps of the invaded islands. Release points on Tahiti marked as dots (after Meyer 1996 and Meyer *et al.* 2012).

introduced to Tahiti and this was approved by the French Polynesian Government. He started up the Miconia Biological Control Programme with introductions of the fungus in 2000 and in 2002. Rather than simply releasing the fungus and then sitting back to watch what happened, as had been done with the *Euglandina* introductions, the releases took place in two permanent monitoring plots. The spread of the fungus in and around these has been followed since then and the changes in other plant species have also been monitored. The aim of this was to be able to show whether or not the fungus was an effective biological control. If it was not, or if there were any unforeseen negative consequences, there would be no way of removing it, but at least the data collected could inform any future plans for release.

The fungus was successfully established at the two sites, infecting all the *Miconia* plants in the monitoring plots and spreading across Tahiti and to Moorea within three years. As anticipated, it caused extensive damage to *Miconia*, resulting in 5-35% defoliation.

The ecological impact of the fungus was evaluated by studying the rare endemic shrub *Ophiorrhiza ubumbellata*. This plant was one of the species discovered by Georg Forster on Cook's second expedition in 1774. The shrubs were found to grow faster and to have higher fertility in the areas where *Miconia* was most defoliated. Encouragingly this led to an increase in the number of seedlings.

The fungus spread by itself to Moorea in 2003 and was deliberately introduced to Raiatea and Tahaa in 2004. It seems that this biological control agent is reducing some of the impact of *Miconia*, although it is probably not a solution in itself. Other approaches of control are still being looked into.

Here the *Partula* story shows the two contrasting faces of biological control. On the one hand there is the old-style approach: the release of poorly selected species with inadequate research and, on the other, the careful study and monitoring of the species before, during and after release. Most importantly, the *Miconia* fungus was unlikely to do any damage to anything other than the target species, being a specialist. *Euglandina* was always a terrible choice: it had little chance of seriously affecting its target and as a generalist predator, was always going to attack non-target species. This generalism makes predators poor choices for biological control as specialist predators will be extremely rare. Predatory species simply cannot afford to evolve to be single-prey specialists; it will always be most advantageous to be able to attack a wider range of prey. This is an obvious point but it has not stopped predators being used in biological control.

A recently notorious case of a predatory species being used in biological control is that of the harlequin ladybird *Harmonia axyridis*. In recent years there has been considerable media attention given to the threat this species poses to insects, including other ladybird species, in North America and Europe. This highly invasive species originated in Asia and was spread from its home deliberately. It was released in North America in 1916 to control aphids and scale insects. This failed but repeated attempts were made to establish it between 1964 and 1982, finally succeeding in 1988. It was also released into South America and Europe in 2001 and South Africa in 2004. It was not introduced to the British isles deliberately but colonised the islands in 2004. In many places its spread has been associated with the decline of native ladybird species, through competition and direct predation. Ironically, biological control of this predator is now being investigated.

Predators are likely to make poor and risky biological control agents because of their lack of specificity. Parasites or pathogens (such as fungi) may be both safer and more effective. While the *Miconia* fungus illustrates a well planned, responsible control programme, *Euglandina* will always be the illustration of how to get everything wrong.

Chapter 11. Back in the zoo

Partula turgida

By the end of 1992 we had confirmed that the situation in the wild was bleak for *Partula*. They seemed to be extinct almost everywhere: Tahiti, Moorea and probably Bora Bora. If I had left any behind on Raiatea it would only have been a handful. Tahaa had not been checked but there was little prospect of anything remaining there; even from the distance of Raiatea's Temehani Plateau I had been able to see that Tahaa no longer held any good *Partula* habitat, and there was no reason to suppose that *Euglandina* would not have been released there. Huahine was probably the only *Euglandina*-free island, and it seemed unlikely that would remain the case for much longer. The captive populations were now effectively the last hope for *Partula* survival.

Captive groups were reorganised once again; with the new influx of wild collected *faba* this species was spread out from London, to Jersey, even though the focus of *Partula* conservation was shifting more and more to the Zoological Society of London. As part of the Zoological Society of London, London Zoo is linked to a major zoological research institution in the form of the Institute of Zoology. At the time one of the research fellows at the Institute was Dr. Georgina Mace (now Professor of Biodiversity and Ecosystems at University College, London). One of her interests is assessing threat status of different species, and she was a leading figure in the development of precise criteria for assigning species to categories in the IUCN Red List of threatened species. The global list of species threatened with extinction was developed out of the Red Data Books and over the past 50 years have grown from published books to annually updated on-line lists. As the Red List expanded it became clear that the categories of threat used in the old Red Data Books were very incomplete, and reflected the interests of the Red Data Book authors, rather than the true status of the world's species. It was not all large charismatic mammals, birds and reptiles (although these dominated); for *Partula* the 1986 Red List included only eight Society Island species (*hebe* and the Moorean species were listed as Endangered, except for *aurantia* which was 'Endangered/Extinct'). In 1987 the extinction of the wild Moorean species had been recognised and the 1988 and 1990 Red Lists included all seven species as Extinct (although in fact all but *exigua* were still alive in captivity).

To expand these into an accurate Red List would necessitate the development of scientific criteria for assessing the status of species. The new criteria were devised to be accurate for all species, whether mammal, bird, plant or snail, and were tested on several different groups. A three-day workshop in 1993 evaluated the status of all *Partula* and during this we tried out the draft Red List criteria. This workshop was a major event in *Partula* conservation, drawing together 33 participants to develop plans for conservation of all the species. As a result of the status review the 1994 Red List increased the number of listed Society Island *Partula* species from eight to 57 (34 Extinct, 8 Critically Endangered, 12 Endangered and 3 Vulnerable), with two Critically Endangered and two Endangered *Samoana*. This was a much more accurate reflection of the true status of the snails. Since then there have been a number of refinements, particularly in taxonomy and much better information. As a result the last re-assessment (2007) has 65 species (49 Extinct, 11 Extinct in the Wild, 4 Critically Endangered, 1 Vulnerable). For *Samoana* there are now 1 Extinct (the questionable '*Samoana' jackieburchi*), 3 Critically Endangered and 1 Data Deficient. It can now be said with confidence that all Society Island partulids are threatened, and 91% are extinct in the wild, with 75% being completely extinct. The great majority of these extinctions occurred in the space of 10 years, which is an unprecedented extinction rate.

In 1994 an opportunity arose to create a small predator-proof reserve on Moorea, where captive-bred snails could be released into areas protected from *Euglandina*. The Moorean snails had now been in captivity for many years and had been living in plastic boxes and glass tanks for several generations. It was possible, or even probable, that over that time they had been selected to be good at living in boxes and may no longer have been adapted to living in trees. It is easy to see that a captive-bred lion may need training to enable it to cope with life on the savannah but it may seem a stretch to think that the snails could face similar problems; it was a very real concern though. If they had been selected to live only in plastic boxes, the whole *Partula* conservation effort could well have been doomed to failure. To test whether or not captive-bred *Partula* would be able to survive in the wild an experimental release was carried out on tropical trees where they could be monitored closely. The trial release took place in the Palm House of Kew Gardens. This remains a unique experiment – the deliberate release of snails into a botanic garden! The snails used were sixth generation descendants of the last *taeniata* collected on Moorea in 1982. Each snail was marked with an individual number and a small fluorescent marker glued to its shell. One of the glass tanks containing 30 adult, 10 subadult and 20 juvenile snails was fixed into a small *Pandanus* tree, left for a while for them to acclimatise and on 9th November the door to the tank was opened. Predictably, their escape from the tank was hardly dramatic, being at snail's pace.

Not only were these snails the first (and probably last) to be released into a botanic garden, they are probably the most intensively monitored snails ever. For the first two weeks they were monitored constantly by teams of two observers taking 6-10 hour shifts. During the day they were easy to keep track of and at night the snails could be located by their fluorescent markers. To start with, the position of each snail was recorded every 30 minutes, which gave 15,000 observations. After the first two weeks the monitoring was reduced but kept up for the next 15 months.

It was a relief to find that the snails did survive outside of their artificial boxes. As time went on they dispersed more and became harder to locate, but after 15 months there were still some adults near the release point. Even more encouragingly 16 babies were found. The observations of their resting sites and the feeding behaviour matched what Bryan Clarke, Jim Murray and Michael Johnson had reported on Moorea 20-30 years earlier. After a year the snails had dispersed no more than three metres. That is hardly a great distance however you measure it, but that was also a good result. When releasing animals into the wild you do not want them to all shoot off in different directions, for then they may never find one another for breeding. That is a problem when reintroducing birds in particular; a great deal of effort has to be spent trying to acclimatise the animals to the release point, but even then some individuals do disappear, never to be seen again. The Palm House release showed that a reintroduction could be a success if the snails were put on suitable trees back on Moorea; all we needed to do was to keep out the predators.

London Zoo now had full responsibility for managing the 'International *Partula* Conservation Programme'. The group of organisations and people involved

Fig. 1. The Palm House at Kew Gardens

in *Partula* conservation was now renamed the 'Pacific Islands Land Snail Group' to reflect broader interests. At the same time an action plan was put together: '*Partula* '94: An Action Plan for the Conservation of the Family Partulidae, by the Pacific Island Land Snail Group'. The review of the status of all 117 Partulidae species planned the expansion of the captive breeding programme from 23 to 53 species and set a population target of 250 adults for each species. This would require at least eight more institutions joining the programme. Surveys were also planned for Huahine and the Marquesas islands.

The 'Operation *Partula* '94' expedition was organised for August 1994, with Dave Clarke and Paul Pearce-Kelly of London Zoo, Bryan Clarke, Jim Murray and their families, and Roger Klocek from the Shedd Aquarium. This was to survey *Euglandina*-free Huahine again, Raiatea and Tahaa and to establish the Moorean reserve. It was hoped that the release of snails into the reserve would establish a safe *Partula* population but, more importantly, it would test the practicality of reintroduction. We needed to know whether barriers would be effective in excluding *Euglandina*; would monitoring and maintenance be cost-effective; and would the captive-bred *Partula* adapt to the wild as well as the Kew experiment indicated.

Raiatea and Tahaa were overrun by *Euglandina* and, as expected, no *Partula* could be found. Sadly the Huahine visit was a great disappointment, for the long-feared arrival of *Euglandina* had taken place. The last remaining *Partula arguta* and *varia* were rescued. The Moorean reserve was more positive though. The Gump Biological Research Station was funding the construction and maintenance of the reserve and the French Polynesian government's Service de l'Economie Rurale had agreed to provide the site. Jim and his son Will travelled to Moorea and constructed the reserve along with the Gump staff in an area of nearly pristine forest. This was as natural as it was possible to get, with an almost complete forest canopy over the reserve. An area of 20x20 metres was surrounded by a 75 cm high barrier of galvanized iron roof sheeting. Obviously snails can climb such fences, so at the base of the barrier a plastic trough was filled with salt and at the top a pair of wires was connected to a 12-volt battery. Any *Euglandina* getting past the salt trough would be shocked by the electric fence and fall outside the reserve. In September 1994 100 adults of each of *suturalis, taeniata* and *tohiveana* were released into the reserve. For a year they were to be monitored weekly by students from the Gump Biological Research Station and CRIOBE. During this time any breaches in the barrier would be repaired.

In 1995 Dave Clarke and Jim Murray led a follow-up expedition, accompanied by two zoo staff members Warren Spencer (Bristol Zoo), Jonathon Cracknell (Chester Zoo) and a genetics researcher, Trevor Coote, who was in the middle of a PhD on partulid genetics and conservation at the University of London. They returned to the reserve on Moorea and also made the first surveys of the

Fig. 2. The 1994 *Partula* reserve, with Trevor Coote at the predator-proof barrier. (photo: P. Pearce-Kelly).

Marquesas Islands. These are the outer islands of French Polynesia, with their own partulids (all *Samoana* species) but apparently no *Achatina* or *Euglandina*. It was important to check whether they really were *Euglandina*-free. It turned out that this was no longer the case, the spread of *Euglandina* was continuing. The expedition was joined by Éric Loève, a biologist living in Tahiti who had found some *Partula* in January 1995 still surviving in the southern peninsula of Tahiti-Iti, these were of two species: *affinis* and *hyalina*.

The situation on Moorea was not good. The monitoring had proved to be problematic and after around six months had become erratic. Breaks in the electric fence had not been repaired and *Euglandina* had immediately taken the opportunity to invade the reserve. There were recent-looking *Partula* shells in the reserve, including juveniles, suggesting that there might still be some survivors. In the end just three live adults were found. The reserve was repaired, *Euglandina* removed and a simpler electric fence installed. Restocking the reserve was delayed by difficulties in arranging permits and logistics for transporting the snails. During this delay the reserve was damaged once more, this time by a storm blowing branches down onto the fence, but this time it was quickly repaired by the Gump staff.

In April-May 1996 more snails were taken out to the reserve on Moorea by Dave Clarke, Sara Goodacre and Trevor Coote. There they were joined by Éric Loève and Dr. Jean-Yves Meyer, a botanist at the French Polynesian government's research department. Meyer has a particular interest in invasive plants and is the expert on the highly invasive *Miconia calvescens*. The team repaired minor damage to the reserve and released more snails into it: 80 of each of *tohiveana, suturalis* and *taeniata*. Before releasing the snails they searched each plant for hiding *Euglandina* and turned over every leaf on the ground to remove every single egg. During this search they found a single live juvenile *Samoana attenuata* inside the reserve. *Samoana* have always been a little different; harder to find, usually higher in the trees and noted by everyone who has found one to be 'stickier' than *Partula*. Maybe some of these properties enabled some to survive. While the reserve was restocked discussions were held with the French Polynesia government and it was understood that a second reserve could be constructed on Tahiti after a year of monitoring the Moorean release.

In November the news was bad again. Heavy rains had washed soil out from under the barrier and vigorous plant growth now hung over the fence. *Euglandina* had breached the barrier, and only one *taeniata* and one *tohiveana* survived, although there were still many *suturalis*. Over the next 18 months attempts were made to solve the maintenance problems but finally, in June 1998, the year experiment was halted. The remaining eight snails were removed. Interestingly these were all found to be wild-born snails. The experiment had confirmed what the Kew experiment had also shown: captive-bred *Partula* could survive and breed in the wild. The reserve design had also been shown to be effective in excluding *Euglandina* but maintenance was a problem, particularly in a forest environment where trees could fall onto the fence or creepers grow over it at any time. In 1997 the French Polynesian government finally declared all partulids to be protected species, but field conservation of *Partula* had reached a dead-end, at least temporarily.

Back in the zoos, in 1995 Jersey now added *radiolata* from Guam, keeping this, *faba* and *mooreana*. Other species were kept in just one place; Bryan Clarke had the only remaining *mirabilis* at Nottingham, and London had the only *dentifera, tristis, hebe, labrusca, turgida* and *arguta*. In total 12 institutions were involved (Nottingham, Virginia, Jersey, London, Poznan, St. Louis, Edinburgh, Chester, Detroit, Martin Mere, Bristol and the Conchological Society). They held 20 species (*mirabilis, mooreana, suturalis, taeniata, rosea, varia, dentifera, faba, dentifera, tristis, hebe, tohiveana, hyalina, nodosa, labrusca, turgida, gibba, radiolata, otaheitana, arguta*). Of these *arguta* was the most precarious, with only four snails. By the end of the year these last four had died and this species joined *exigua* and *aurantia* in extinction.

Sadly *arguta* was not the last species to be lost. In 1994 there were 296 *turgida* in captivity and the species seemed to be thriving, but then went into a sudden decline. Over a period of 21 months the population crashed to a single individual. At 15:30 on 1st January 1996 the last *turgida* died, the most precise moment of extinction ever recorded. The last *turgida* was autopsied and, for once, a cause of death was identified. It seems this animal was killed by an infection of the microsporidian parasite *Steinhausia*. This makes it the only species extinction known to have been caused by disease. No other *Partula* autopsies have identified any clear causes of death and *Steinhausia* does not seem to have caused any other problems. Where it came from and why it was so devastating to *turgida* will remain unknown. In 2002 the last *labrusca* died. Although this species had been in captivity for 10 years it had never really become established and its extinction was disappointing but not surprising. It was a high-altitude species and no high-altitude *Partula* or *Samoana* has adapted well to captivity.

London Zoo's *Partula* were moved from the 1913 Insect House to a modern building in 1999. The Millennium Conservation Centre housed the 'Web of Life' display, showcasing the diversity of life, now known as 'BUGS' which stands for Biodiversity Underpinning Global Survival. Dave Clarke managed the Web of Life as its Head Keeper. As in the old Insect House there was a dedicated *Partula* room in the new building, but this time the Snail Room had been designed for the snails. Instead of cramped corners with densely packed shelves of small snail boxes the new room was more spacious, with room for larger, more practical snail tanks and space to make maintenance much easier. It was also controlled for temperature and humidity. An additional novelty was to have the room visible from the public areas, so that visitors to the zoo could see the day to day work of conserving the snails.

All of the Society Islands were surveyed once more in 2000 by Trevor Coote and several Tahiti residents, including Éric Loève. By this time *Euglandina* had appeared on Huahine and wiped out the three *Partula* species living there (*rosea*, *varia* and *arguta*). The survey confirmed the extinction of *Partula* on Raiatea, Bora Bora and Tahaa, removing any hope of the survival of the two species that had been lost in captivity by that time: *arguta* and *turgida*.

At this time Jersey Zoo (now renamed the Durrell Wildlife Conservation Centre) was consolidating its conservation projects. Its *Partula* had never really thrived and in 2004 they gave up on keeping *radiolata*. In 2009 the declining *faba* were sent to Bristol, where all the surviving animals of this struggling species were consolidated. They also stopped keeping *mooreana* and, finally, in 2010 Jersey sent their *taeniata* to London, leaving the *Partula* programme after 29 years.

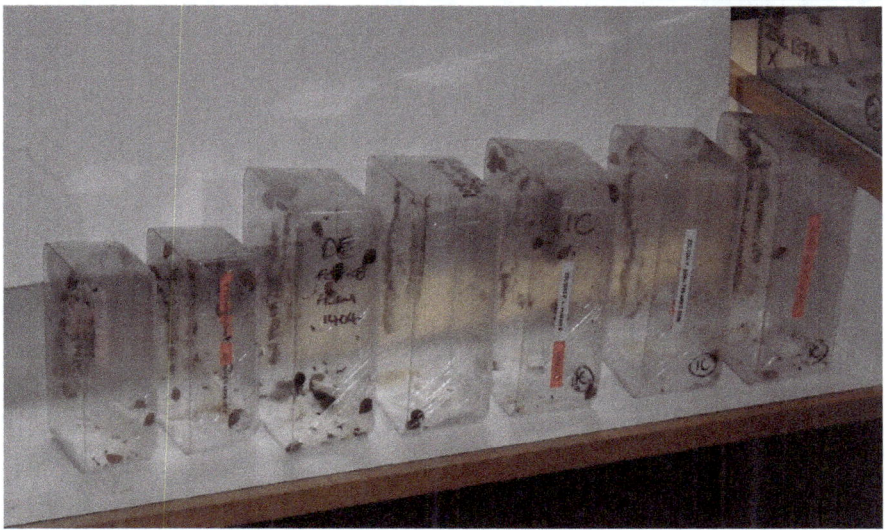

Fig. 3. The London Zoo snail room viewed from the public gallery (top), and *Partula* boxes (bottom)

Despite the loss of Jersey, the *Partula* programme was not declining; in 2008 Copenhagen Zoo had joined in. At the start of 2013 the *Partula* programme kept 17,721 snails. For stability the aim is to keep 250 adults of each species. Nine species were just above that level and *tristis* was very close at 228. Some had become super-abundant: there were 680 adult *dentifera* and an impressive 1,324 *nodosa*. Others struggled; for example, even though it was the most widespread and adaptable *Partula* there were just 57 captive adult *hyalina* in 2013 Most worryingly *faba* was still not recovering, with just 15 adults remaining.

Chapter 12. Rediscoveries

Partula affinis

Despite appearances, not all *Partula* had succumbed to *Euglandina*. In 1992 Bryan Clarke had told me of the finding of some *otaheitana* high up on Mount Marau on Tahiti. The next surprise, also on Tahiti, was in 1995 when a few populations of *affinis* were found, suggesting that there was a need for more surveys.

In 1994 Paul Pearce-Kelly investigated the *otaheitana* population on Mount Marau using the unusual approach of taking a taxi up to the viewpoint. He asked the mystified taxi driver to stop every now and then and had a quick look at the trees in the ravines below the road. Not surprisingly there were *Euglandina* everywhere, but just below the viewpoint, at 1,300 metres above sea level, he found the surviving population of *otaheitana*. This seemed to be just above the altitude limit for *Euglandina* presence. In 1996 *Samoana burchi* was found there as well, both species extending down to 1,000 metres. The first *Euglandina* was found just 100 metres below that. These populations persist to this day; sometimes *Euglandina* moves up a little further (on one occasion up to 1,420 metres) and then the partulids suffer, but the invasion never seems to be permanent. This appears to be the very edge of the altitude ceiling that I had predicted in 1994. It is a little higher than I predicted, but it still seems to be there.

On Tahiti small surviving populations of a few species were found between 1995-7 in the interior of the main island of Tahiti Nui. This is difficult to access and these survivors had been overlooked before. On the smaller Tahiti Iti the best populations survived in the coastal south-east, most notably in Faaroa Valley which was occupied by *clara, affinis, otaheitana* and *hyalina*, along with *Samoana attenuata* until these were invaded by *Euglandina* in 2001. Then *affinis* and *otaheitana* vanished, and the others became very rare.

In 2003 field conservation ideas were revitalised when an agreement was reached between the French Polynesian Government and the International Partulid Conservation Programme. Since then Trevor Coote has been based in Tahiti as the IPCP field-biologist funded by the *Partula*-keeping zoos. He has carried out more detailed field surveys in the Society Islands, the Marquesas and the Australs, and monitors the Tahitian and Moorean populations, covering 20 sites in 13 valleys and one high altitude location. New plans have been devised for restoring the 10 species that are extinct in the wild but remain in captivity.

Trevor Coote's activity on Tahiti immediately produced results. In 2004 he searched 69 valleys and in 19 found small populations of *clara* and *hyalina*. More populations have turned up since then and they are now known from 33 valleys. In none are they common; in many they are known from a single sighting, and that may not be in every year. However, at least the odd individual of *clara* and *hyalina* seem to be able to survive with *Euglandina* around. They are lowland species, so the altitude ceiling does not save them; they must have some characteristic that enables them to survive better than all the other species. Crampton had earlier noted that these two species tended to carry three or four embryos instead of the more usual two or three, suggesting a higher reproductive rate, but even that is far too slow to cope with *Euglandina's* voracious appetite. Crampton also commented that they were tolerant of dry areas and 'more active' than other *Partula*. As other *Partula* are almost completely inactive most of the time, this hardly gives the levels of speed needed to escape a hungry *Euglandina*. We still have no sensible explanation for how they survive.

In the high areas the *otaheitana* on Mt. Marau were facing new threats, with clearance of some patches of forest for agricultural plots (despite the steep terrain) and a major clearance to build a radar station for the airport. Despite this the snails were surviving; getting access to the sites was harder than finding the snails once there. There were some *Euglandina* there from time to time and in 2005 a *Euglandina* population explosion completely cleared one ravine at 967 metres.

2005 was also a good year for discoveries on Tahiti with more *clara* and *hyalina* and the rediscovery of living *affinis*. More high altitude *otaheitana* have been found, mostly above the *Euglandina* populations. Three species of *Samoana* have also been rediscovered on Tahiti: high altitude populations of *burchi* and *diaphana*, as well as the more widespread *attenuata*.

Down in the south-east of Tahiti, on the peninsula of Tahiti-Iti, a remarkable four *Partula* species had been recorded in 1995. This was a wild area, with no direct road access and difficult terrain. Here there were *otaheitana, affinis, clara* and *hyalina*, as well as *Samoana attenuata*. There were some *Euglandina* but they were localised; perhaps they had never really established themselves there. By 2001 the *Partula* were declining and *otaheitana* seemed to have disappeared. A reserve was built to safeguard the remaining species but this was destroyed by rockfall and the attempt to protect them abandoned due to the difficulty of reaching the area. The decline continued and by 2011 the *affinis* seemed to have vanished and *hyalina* and *clara* had become scarce. Strangely the following year two volunteers, Jeff and Karen Lambert spent their honeymoon looking for *Partula* and relocated the *affinis* population.

On Moorea it seemed unlikely that any viable populations survived. A *taeniata* turned up in the Kellum Gardens in Opunohu Bay in 1994 and had been found there again in 1995 but had vanished by the following year. A survey in 2000 did again encounter the elusive *Samoana attenuata*, only ever found on Moorea as a rare isolated record. This survey also found *taeniata* in six locations, but none of these appeared to be significant populations. However, an apparently viable *taeniata* population was discovered in 2002 in mangrove ferns at Opunohu by Carole Hickman from the University of Berkley, working at the Gump station. This was particularly surprising as no *Partula* had ever been recorded in such habitat before. It remains a unique population: the only *Partula* to live on plants that are surrounded by sea-water at high tide. In Moruu valley a small population of banded and white *taeniata* were found in early 2005. *Euglandina* had cleaned out the area in the late 1980s but must have missed some *Partula*, and had itself since disappeared. No sooner had the *Partula* been rediscovered in Moruu than *Euglandina* reappeared; the *Partula* declined, but a few still persist. Strangely though the banded snails have all vanished, and only white ones remain.

Further Moorean survivors were found in other valleys from 2005, but these all declined again, sometimes disappearing completely. All this suggests that *taeniata* and *Samoana* persist on Moorea, but are extremely scarce. At these numbers it seems unlikely that they can survive indefinitely. A possible exception is Maramu valley. This was one of the areas that Jim and Bess Murray had searched in 1987 and had found strewn with empty shells but devoid of live *Partula*, but in 2009 numerous *taeniata* were found at one small location. They have been found there regularly since and at least 100 were counted in 2013, the site now has one of the largest *Partula* populations remaining in the islands. This population seems to be expanding in numbers and area, returning to something like its original state in the absence of *Euglandina*.

Huahine lost all three of its *Partula* in about 1996, but there has been a sighting of a *Samoana* on its mountain. Mount Turi is not high enough for an altitude ceiling to limit *Euglandina*, so this is not a population in a predator-free refuge. It is probably just *Samoana*'s unexplained ability to hang on long after *Partula* have been eaten to extinction.

On Raiatea there was little hope of finding survivors. I had collected all I could in 1992, *Euglandina* had not reached the highest parts of the island then, but the Temehani Plateau where *labrusca* and *faba* survived had been invaded by *Euglandina* in about 1996: Raiatea's altitude ceiling had been breached. Then, in 2006, Jean-Yves Meyer organised a botanical expedition to the top of the island with interesting results. Mount Toomaru (or Tefatua) had been climbed once before, by another botanist, John W. Moore in 1927. Moore's nine-month collecting trip to Raiatea resulted in a collection of many interesting plants and a great many snails, the latter remain unidentified. There are no records of what must have been an adventurous climb of Mount Toomaru. When Meyer ascended the mountain it took three days to cut a path to the top at 1017 metres. There they surveyed the plants and while doing so Meyer spotted a *Partula* on a leaf. He found two more on the ground and collected one of them for identification. This turned out to be a new species, now named *Partula meyeri*. He also observed a *Samoana attenuata*. This was not so surprising, but the finding of a new and living *Partula* was remarkable. Subsequently a *Samoana* has also been found on the Temehani Plateau.

With the greatly improved communication between the French Polynesian government and the Partulid Global Species Management Programme that Trevor Coote has managed to establish, thought turned to trying the reserves once more. In 2004 the best choice seemed to be a reserve in Papehue valley on Tahiti to reintroduce some of the thriving captive Tahitian snails. Progress on finding a good location was slow and in 2007 the government offered to provide a site in the Te Faaiti Natural Park in the centre of the island. A year later a final site had been selected on a small plateau in woodland infested with introduced plants. A future site in the Faaroa Valley on Raiatea was also selected. In the Te Faaiti Natural Park there was a project to manage the habitat of the area by removing invasive species over a five hectare site. This would allow the reserve to be placed, not in closed forest as had been the case on Moorea, but in a clearing. This meant that storms would not cause falling trees to smash the barriers, but the reserve would require planting to create habitat. The reserve project did not make smooth progress, this time not for ecological reasons but because of the wider context. The global financial crisis starting in 2008 seriously affected government budgets and they were forced to withdraw their financial support. Fortunately the funding agency, the Critical Ecosystems Partnership Fund was willing for funds to be reallocated within the habitat restoration project and the 12x9 m reserve was built in 2012. Releases into this were planned for 2013 but government reorganisations meant that they had to be delayed until 2014. This will include *nodosa* as this is the most successful species in captivity at present. However, its natural distribution is in the dry west coastal part of Tahiti and the release site is in the wetter interior. As a comparison a release will also be made into its natural range at Papehue Valley. Due to land ownership issues this will not be a predator-proof enclosures but currently *Euglandina* are absent from the area. *affinis* will also be included in the Te Faaiti reserve, along with *hyalina* if stock levels permit.

Fig. 1. *Partula meyeri* and its habitat on Mt. Toomaru. Top – view of Mt. Toomaru (left-hand peak) from the east. Bottom – the *Partula meyeri* locality. Photos of habitat and *Partula meyeri* by Jean-Yves Meyer, Délégation à la Recherche de la Polynésie française, Tahiti

Chapter 13. What does the future hold?

Partula meyeri

At the time of writing the picture for *Partula* survival is decidedly mixed. In the wild some species cling on somehow, but new threats are added to the existing ones. Some of these threats are variants of old ones, for *Euglandina* is not the only alien predator *Partula* face. As *Euglandina* proved not to be the cure-all that had been hoped, the search for more effective predators continued and a new biological control agent was tried. This goes back to the early days of *Euglandina* experimentation; when H.J. de Wilde de Ligny observed the disappearance of snails in New Guinea in areas occupied by a flatworm, what he was observing was the impact of *Platydemus manokwari* on the snails.

There were a few records of *Platydemus* from New Guinea in 1962-73, and then it was found in Australia in 1976 (Queensland, then Northern Territory in 2002). Next it appeared in Guam in 1977, presumably as a result of the accidental introduction of flatworms hidden in plants imported from east Asia. By 1981 it was reported that *Achatina* had declined by 95% and at the same time *Platydemus* had appeared on neighbouring Saipan island, then Tinian in 1984, Rota in 1988 and Aguiguan by 1992. The decline of *Achatina* led to the deliberate movement of *Platydemus* from Guam to the Philippines in December 1981 and February 1982. 150 worms were released on Bugsuk Island and in September 1983 *Achatina* was reported to have declined there as well. Dr. Rangaswamy Muniappan, was then the Associate Director of the Guam Agricultural Experimental Station, University of Guam (now the Director Integrated Pest Management in the Collaborative Research Support Program at VirginiaTech) and his observations of the situation on Guam made him the main proponent of the effectiveness of *Platydemus* and he encouraged its release in the Maldives (1985) and elsewhere. It has since been recorded in Micronesia (1983), Japan (Yohohama 1984, Okinawa by 1990, Bonin islands early 1990s), Palau (1991), Hawaii (by 1992), Samoa (by 2000), Vanuatu (2002), Tonga (by 2002) and Fiji (by 2012).

Fig. 1. *Platydemus manokwari*, photographed at Fare Rau Ape on Tahiti, 2013, by Jean-Yves Meyer, Délégation à la Recherche de la Polynésie française, Tahiti.

Euglandina and *Platydemus* received some attention in 1994 with the broadcast of a documentary on them by the BBC. "Predator" highlighted the demise of *Partula* due to *Euglandina* and raised the fear that *Platydemus* could prove at least as bad and cited Dr. Muniappan as being responsible for the spread of the worm. Muniappan took exception to extinctions be attributed to his actions and sued the BBC which eventually settled out of court, paying his expenses and an undisclosed settlement. The BBC's copies of the film were destroyed.

The impacts of *Platydemus* are far from clear. It has been studied on the Bonin islands of Japan and there is convincing evidence from there that it can be devastating for land snails. Just like *Euglandina*, it will follow slime trails up into the trees, so may well prove a problem for any *Partula* that have so far escaped *Euglandina*. Guam has the longest history of both *Euglandina* and *Platydemus*: *Euglandina* introduced in 1957 and *Platydemus* in 1977. By 1989 *Partula gibba* was only easy to find where the flatworms were still absent. Of the other islands, Agiguan was the first target for biological control in the early experimental days of the '*Gonaxis* project'. The *Gonaxis* experiment on Aguiguan was effectively abandoned after Davis had concluded that it had been a great success in 1962. The situation was not re-examined for a further two decades. In 1984 a research team from the University of Guam were unable to find any living *Achatina* or *Gonaxis* on Agiguan. No *Partula* were noted at the time but the following year 14 living *Partula* (*langfordi* and *gibba*) were found, representing tiny relicts of the substantial populations that Kondo had recorded in 1955. In 1995 Barry Smith of the University of Guam found abundant dead shells of many snail species, including *Partula* and *Gonaxis*, but no living *Partula* could be found. He also found the predatory flatworm for the first time. This seems to have been introduced to Aguiguan from Tinian (where it had appeared in 1984) in 1992 as stow-aways in plants imported for a reforestation project. The timing of these observations suggests that the flatworms were responsible for the mass mortality of all snail species on Agiguan.

Partula have also gone from Rota and Tinian, where *Platydemus* are present, but they are still common on some of the more isolated northern islands such as Pagan and Sarigan which remain predator-free. Even on little, isolated, predator-free Pagan island *Partula* faces an uncertain future. Pagan has had a tumultuous history: settled first by Chamorro Islanders then colonised by Spain, sold to the German Empire as a coconut plantation for the German-Japanese 'Pagan Society', devastated by typhoons, captured by Japan in World War I, and then American territory after the defeat of the Japanese Empire in World War II. It is now abandoned following volcanic eruptions in the 1980s. Its abandonment may have saved it from the introduction of predators but there are other threats. A Japanese investment group considered using Pagan as a dump site for debris from the 2011 tsunami. No sooner was that idea dropped than the US military decided to turn the island into a live-fire training base. Other such training bases in the Pacific (in Hawaii and the Marianas) have resulted in completely devastated wastelands – certainly not good for *Partula*.

In Polynesia *Platydemus* appeared on isolated Mangareva island in 1997 and on Moorea in 2009. It has since been confirmed from the Tiapa Valley on Tahiti, and very recently on Mount Marau. At the moment *Euglandina* remains the big problem for Society Island partulids. *Platydemus* may be even more of a threat in the future, but it is a generalist predator that will feed on *Partula*, earthworms, insects, *Achatina* and *Euglandina*. Who knows what impacts it will have on *Euglandina* in the future? All we know for certain is that with the introduction of *Platydemus* the ecosystems of the islands have been destabilised once again.

There is a footnote to the *Platydemus* story; in 2013 the flatworm appeared in yet another new locality, this time in France. The flatworm was found in a greenhouse in the Jardin des Plantes of Caen in the south of France. It was presumably imported as a stow-away along with some plants. Most of France may be too cold for the species to survive over winter (it requires temperatures above 15° C for breeding and 10° C for juvenile survival), but the south of France may well be suitable.

Not all partulids have declined because of predation; in the Marquesas where *Euglandina* is patchy most *Samoana* (*Partula* is not found in the Marquesas) had declined. Where predators are present *Samoana* has been restricted to the higher parts of the islands. On Ua Pou *bellula* is restricted to the high areas as these retain the only remaining tree-snail habitat. Notably, Ua Huka lacks predators and has good habitat, so should be ideal for *Samoana*, but *strigata* is only found in the higher areas. Similarly on Hiva Oa *ganymedes* was found in 1995 but seemed to have vanished by 2011 for no apparent reason. On Raivave island in the Austral Islands *Samoana dryas*, *hamadryas* and *oreas* survive on the mountain, but not in the lowlands. As there are no predators in these areas and the habitat remains intact, the loss of lowland *Samoana* is suspected to be due to that most current of all threats: climate change.

Climate change has been extensively discussed over the past three or four decades, but is not a new idea. As early as 1894 the Swedish chemist Svante Arrhenius realised that the carbon dioxide released by burning of fossil fuels would have an impact on the climate by trapping heat in the upper atmosphere. The term 'greenhouse effect' dates back to 1917 when the inventor Alexander Graham Bell referred to a "sort of greenhouse effect" arising from burning fossil fuels and feared that if unchecked "the greenhouse becomes a sort of hot-house". One of the difficulties in evaluating the importance of Arrhenius's idea was that the concentration of carbon dioxide in the atmosphere was unknown. In 1938 the British engineer Guy Stewart Callendar provided some evidence that carbon dioxide levels and temperatures had risen since Arrhenius's time, but the data were too poor to be fully convincing. It was not until 1958 that carbon dioxide concentrations were monitored consistently. In 1958 carbon dioxide levels were 315 parts per million which had risen to 397 in 2013. The extraction of ancient carbon dioxide in gas bubbles frozen into ice cores has since enabled us to determine that the 2013 level is 39-44% higher than pre-industrialisation levels.

The earth's climate is driven by the energy from the sun. This provides the heat we feel directly but also affects the circulation of air and moisture on land and of water in the oceans. Of the solar radiation arriving on earth, around half is absorbed at the earth's surface and half reflected back to the atmosphere. There it is lost into space or is re-radiated back to the earth. Carbon dioxide and other greenhouse gases absorb some of this infrared radiation, trapping the heat in the atmosphere. This trapping of energy in the atmosphere is Alexander Graham Bell's 'greenhouse effect', resulting in generally increasing air temperatures and changed patterns of air and water circulation. This much had been predicted by Arrhenius, but making actualluseful predictions is a different matter. Climate models are staggeringly complex, but contain too many imprecisions to allow more than short-term forecasting. This means that climate change projections have inevitably given a range of scenarios rather than specific predictions. This uncertainty has frequently, but mistakenly, been interpreted as uncertainty over the science. However these issues are presented, the facts of changing climates can no longer be disregarded. By the end of 2013 global average temperatures had risen by 0.85°C since Arrhenius suggested the possibility. This global figure is pretty meaningless as local factors dominate actual climates and temperature change varies locally, making the popular term 'global warming' unfortunately misleading. Rainfall patterns are also changing and extreme weather events increasing in severity as would be expected with increased energy in the system.

When climate change is discussed with respect to islands most attention is given to sea level rise. Although sea level at Tahiti has risen by more than 7.5cm over the past 30 years, this is of little importance to most *Partula*. The only exception is the one *taeniata* population found at sea level on Moorea. For most populations it is temperature and rainfall that are likely to be the most critical factors.

In Polynesia temperatures have been rising at a rate of 0.39°C per decade, and this may be a significant stress on the *Samoana* species, which are almost exclusively found at cool, high altitudes. Climate change is likely to be having an impact on *Partula* as well. Lowland areas are becoming hotter and dryer, which may affect the relict *clara, hyalina, affinis* and *taeniata* and may also influence the success of future reintroductions. At higher altitudes the greatest concern may be that rising temperatures may make it easier for *Euglandina* to invade the last areas where *Partula* remain relatively abundant. *Partula* survival and extinction has thrown up many unexpected surprises; now we must find a way of predicting the interactions between *Partula, Euglandina* and flatworms in a time of climate change.

In addition to direct effects of temperature on snails and their predators, climate change is likely to have notable impacts on habitats. Changes to habitat are also a serious issue in a changing climate. From my own experience I have seen high forest ecosystems change rapidly in tropical islands as dry seasons have become longer. Cloud forests that were only exposed to an occasional few hours of sun were unable to adapt to days without cloud. As a result a normally water saturated environment became dry, and spongy moss turned to dust. Animals that depended on a wet environment, like frogs and snails, declined markedly. Species that were naturally restricted to the very wettest cloud forests declined to extinction within the space of five years. This shows just how rapid extinction can be in small, isolated populations. Tahiti has larger cloud forests and may not be quite so vulnerable, but the tiny cloud forest on Raiatea is very precarious.

The temperature rise in Polynesia is expected to cause a decrease in cloud forest habitat on Tahiti of 90% by 2100, shifting the lower altitude limit of the important cloud forests by more than 5 metres a year. This would dramatically reduce the habitat on Tahiti occupied by *Partula otaheitana* and *Samoana* species. On Raiatea when *Partula meyeri* was discovered in 2006 it was already restricted to areas above 950 m; this species could lose all of its habitat within just 12 years of discovery, or by 2017.

Partula are clearly facing terrible threats, but why should we care? *Partula* is historically interesting, from the great explorers through to the development of evolutionary biology. They used to be of some cultural significance, through their use in Polynesian jewellery. Cultural, historical and ethical considerations aside, does the loss of *Partula* matter? The honest answer is that we do not know. We do know, however, that *Partula* used to be astonishingly abundant. Garrett's anecdotal observations indicate that, as do Crampton's collections. More precisely, the three Professors recorded an average of 104 snails in 100m^2. Recent dissections of their preserved *Partula* from the 1960s show that they were not herbivores, but grazed on rotting leaves, algae and lichen, with some species specialising on moulds and mildews. No-one has studied the feeding rate of *Partula*, but using a rough figure from studies of other snails they may consume as much as a third of their body weight a day. This would suggest that *Partula* removal of algae and fungi from plant leaves would be in the region of 40 grammes a day per metre. So I would conclude that the loss of *Partula* does matter. With their extinction this cleaning is no longer being undertaken, making the spread

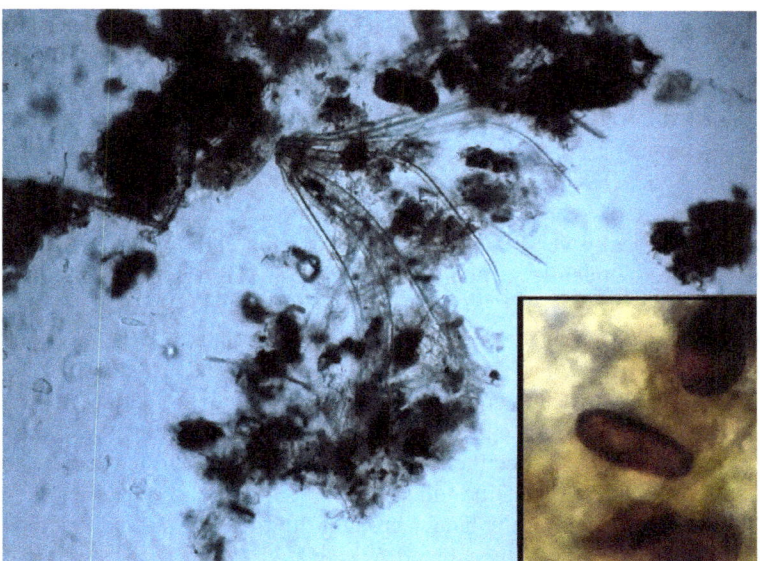

Fig. 2. *Partula* gut contents, showing a variety of plant matter, including the distinctive hairs of *Hibiscus tiliaceus*. Inset: mildew spores.

of plant diseases more likely and reducing the health of some plant species, making invasion by alien species even easier than is was before.

After 17 years of finding relict *Partula* populations on Tahiti, only to see *Euglandina* wipe them out, Trevor Coote thought he had seen everything. A surprise came in 2012 when he found some *clara* in a valley on the west of the island. He had seen these five years earlier, in a patch of wild ginger, but a few months later *Euglandina* had found them. For five years the population had been thought to be extinct, but they were still there. The reason they had remained hidden, both from *Euglandina* and from anyone searching for them, was that the population was not actually in the low-growing, easily accessible ginger, but in the tall Tahitian chestnut or 'mape' trees, *Inocarpus fragifer*, growing above the ginger. It is probable that those found in the ginger had simply fallen out of the mape tree. So far the tree population has remained untouched by *Euglandina* and it seems that the predators are not keen to climb up the rather dry mape trunks. This offers a small glimmer of hope: maybe the *clara* and *hyalina* that live on mape today might survive there indefinitely, and maybe other species could be induced to live on the trees. If other *Partula* can be reintroduced to the relative safety of mape trees it might just enable *Partula* to start to adapt to the presence of *Euglandina*.

It is clear that the 17 remaining Polynesian *Partula* face great uncertainty in the wild: devastated by *Euglandina* and now facing probable threats from flatworms and climate change. Many of the species that survive in captivity are stable or thriving, but not all. At the moment the most vulnerable species of all is *Partula faba*. Once the commonest *Partula* on Raiatea, *faba* was still abundant where it survived in 1991. When I collected the last survivors in 1992 it was restricted to just two sites, but was plentiful there. Everything pointed to this species being easy to keep: it was a large snail, historically very widespread and not apparently a specialist. Despite this it never thrived. Juvenile survival has never been very good and although the number of adults rose to around 180 in 1995 it has declined steadily since then. In April 2010 there were only three remaining captive groups at Bristol and Jersey zoos and Imperial College. The species was at crisis point so the snails at Jersey and Imperial were sent to Bristol to create a single consolidated population, numbering just eight adults. There seemed to be some progress, for at the end of 2012 there were 15 adults and 24 juveniles. Since then though there has been further decline and in mid-2014 there remained just 15 individuals.

245 years after Daniel Solander brought the first *Partula* out of the South Pacific that same species hangs by a thread. With just 15 *Partula faba* surviving, this is a symbol of our impacts on the world around us, but also of our responsibility to turn back the declines.

References

Chapter 1
Forster, G.A. 1777. *Voyage around the World in His Britannic Majesty's Sloop Resolution.* London

Gilbert, L.A. 1966. Banks, Sir Joseph (1743–1820). *Australian Dictionary of Biography,* **1**: 52–55

Hawkesworth, J. 1773. *An account of the voyages undertaken by the order of his present Majesty for making discoveries in the Southern Hemisphere, and successively performed by Commodore Byron, Captain Wallis, Captain Carteret, and Captain Cook, in the Dolphin, the Swallow, and the Endeavour: drawn up from the journals which were kept by the several commanders, and from the papers of Joseph Banks esq.* Strahan & Cadell, London.

Rauschenberg, R.A. 1968. Daniel Carl Solander: Naturalist on the "Endeavour". *Trans. Amer. Phil. Soc.* N.S. **58**(8): 1-66

Chapter 2
Broderip, W.J. 1832. New species of shells collected by Mr. Cuming on the western coast of South America and in the islands of the South Pacific Ocean. *Proc. Comm. Sci. Corresp. Zool. Soc. Lond.* **1832**(2): 124-126.

Bruguière, J.G. [ed.] & C.H. Hwass. 1792. *Encyclopédie méthodique. Histoire naturelle des vers. Tome première partie 2.* Panckoucke, Paris.

Chemnitz, J.H. 1786. *Neues systematisches Conchylien-Cabinet / geordnet und beschrieben von Friedrich Heinrich Wilhelm Martini und unter dessen Aufsicht nach der Natur gezeichnet und mit lebendigen Farben erleuchtet.* Vol ix.

Chenu, J.C. 1845. *Bibliothèque conchyliologique.* A. Franck, Paris.

Dall, W. H. 1905. Thomas Martyn and the Universal Conchologist. *Proc. U.S. nat. Mus.* **19**: 415-432

Dance, S.P. 1980. Hugh Cuming (1791-1865) Prince of collectors. *J. Soc. Biblphy. Nat. Hist.* **9**(4): 477-501

Dance, P. 1986. *A history of shell collecting.* E.J. Brill - Dr. W. Backhuys, Leiden.

Dance, S.P. (undated) *Delights for the eyes and the mind. A brief survey of conchological books.* www.bionica.info/biblioteca/DanceBibliophile.pdf

Deshayes, G.P. in Lamarck. 1816. *Histoire Naturelle des Animaux sans Vertebres.* 2(VIII)

Fabricius, J.C. 1805. Johan Christian Fabricius, Professor ved Universitetet I Kiel. In: Lahde, G.L. (ed.) *Portaetter med Biographier af Danske, Norske og*

Holsteenere. Andreas Seidelin, Copenhagen.

Férussac, A.E.J.P.J.F. d'Audebard de 1821-1822. *Tableaux systématiques des animaux mollusques classés en familles naturelles, dans lesquels on a établi la concordance de tous les systèmes; suivis d'un prodrome général pour tous les mollusques terrestres ou fluviatiles, vivants ou fossiles*. Paris

Garrett, A. 1884. The terrestrial Mollusca inhabiting the Society Islands. *Journ. Acad. Sci. Phil.* (2) **9**: 17-114

Gmelin, J.F. 1791. *Systema naturae 13th edition, volume 1, part 6: Vermes*. Emanual Beer, Leipzig.

Gray, J.E. 1868. Notes on the specimens of Calyptraeidae in Mr. Cuming's collection. *Proc. Zool. Soc. Lond.* (**1867**): 726–748

Huish, R. (ed.) 1839. *A Narrative of the Voyages and Travels of Captain Beechey To the Pacific.* London

ICZN. 1957. Opinion 456: Rejection of the works by Thomas Martyn published in 1784 with the title "the universal conchologist" as a work, which does not comply with the requirements of article 25 of the "règles" and which therefore possesses no status in zoological nomenclature and rejection also of a proposal that the foregoing work should be validated under the plenary powers. *ICZN Opinions and Declarations* **15**: 393-418

Kohn, A.J. 2009. Type specimens of *Conus* (Mollusca: Gastropoda) in the Zoological Museum of the University of Copenhagen: A Historical Account. *Steenstrupia* **30**(2): 97–113.

Lightfoot, J. 1786. *A catalogue of the Portland Museum: lately the property of the Duchess Dowager of Portland deceased Which will be Sold by Auction, by Mr. Skinner and Co. on Monday the 24th of April, 1786, and the thirty-seven following days, At Twelve O'Clock, Sundays, and the 5th of June, (the Day his Majesty's Birth-Day is kept) excepted. At the late Dwelling-House, In Privy-Garden, Whitehall; By Order of the Acting Executrix.* (Skinner), London.

Martyn, T. 1784. *Universal Conchologist, exhibiting the figure of every known shell accurately drawn and painted after nature with a new systematic arrangement.* London.

Martyn, T. 1789. *Figures of non descript shells, collected in the different voyages to the South Seas since the year 1764. Published by Thomas Martyn, and sold at his house, No. 16, Great Marlborough Street, London.* The Universal Conchologist [2n ed.]. Pp. 1-39, explanatory table, 80 pls.

Reeve, L.A. 1850. Monograph of the genus *Partula*. *Conchologia Iconica* **6**: 9 pp, 4 pls.

Chapter 3
Garrett, A. 1874. Fische der Südsee beschrieben und redigirt von Albert C.L.G. Günther Heft V. *Journal des Museum Godeffroy* Heft II. Hamburg, L. Friedrichsen & Co.
Green, K.W. 1960. William Harper Pease – his 21 years in Hawaii. *Hawaiian Shell News* (**1960**): 5
Johnson, R.I. 1995. Types of shelled Indo-Pacific Mollusca described by W.H. Pease. *Bull. Mus. Comp. Zool.* **154**:1-61
McMichael, D.F. 1969. Brazier, John William (1842–1930). *Australian Dictionary of Biography* **3**. Reprinted online: http://adb.anu.edu.au/biography/ brazier-john-william-3046/text4479 [accessed 18 August 2013]
Pease, W.H. 1864. Descriptions of new species of land snails from the islands of the central Pacific. *Proc. Zool. Soc. Lond.* (**1864**): 668-676
Thomas, W.S. 1979. A biography of Andrew Garrett, early naturalist in Polynesia: part 1. *The Nautilus* **93**(1): 15-28

Chapter 4
Boycott, A.E., C. Diver, S.L. Garstang & F.M. Turner. 1930. The inheritance of sinistrality in *Limnaea peregra* (Mollusca: Pulmonata). *Phil. Trans. Roy. Soc. London* B **219**: 51-131
Cheminee, J.L., R. Hekinian, J. Talandier, F. Albarede, C. W. Devey, J. Francheteau & Y. Lancelot. 1989. Geology of an active hot spot: Teahitia-Mehetia region in the South Central Pacific. *Marine Geophys. Res.* **11**(1): 27-50
Darwin Correspondence Database, https://www.darwinproject.ac.uk/entry-4989 and https://darwinproject.ac.uk/entry-5260 [accessed on 16 September 2013]
Gulick, A. 1932. *Evolutionist and missionary. - John Thomas Gulick: Portrayed through documents and discussions*. University of Chicago Press, Chicago
Gulick, J.T. 1888. Divergent evolution through cumulative segregation. *Journ. Linn. Soc. Lond., Zool.* **20**(120): 189–274
Hall, B.K. 2006. "Evolutionist and missionary," The Reverend John Thomas Gulick (1832-1923). Part 1: cumulative segregation – geographical isolation. *J. Exp. Zool. B. Mol. Dev. Evol.* **306**: 407-418
Kohler, R.E. 2002. Landscapes and Labscapes. Exploring the Lab–Field. Border in Biology. University of Chicago Press, Chicago
Kondo, Y. 1980. *Samoana jackieburchi*, new species (Gastropoda: Pulmonata: Partulidae). *Malacol. Rev.* **13**: 25–32
Kondo, Y. 1955. *A revision of the family Partulidae*. PhD thesis, unpublished. Harvard University.
Kondo, Y. 1956. Henry Edward Crampton, 1876–1956. *Nautilus* **70**: 31-34
Mendel, G. 1950. Gregor Mendel's Letters to Carl Nägeli. *Genetics* **35**(5, pt 2): 1–157

29. Online as http://www.esp.org/foundations/genetics/classical/holdings/m/gm-let.pdf

Moore, R. 2001. The "Rediscovery" of Mendel's Work. *Bioscene* **27**: 13-24

Romanes, G. 1897. *Darwin and After Darwin.* Vol. **3**. The Open Court Publishing Company

Stamhuis, I.H., O.G. Meijer & E.J.A. Zevenhuizen. 1999. Hugo de Vries on Heredity, 1889-1903: Statistics, Mendelian Laws, Pangenes. *Isis* **90**(2): 238-267

Sturtevant, A.H. 1923. Inheritance of direction of coiling in *Limnaea. Science* **58**: 269-270

Wright, S. 1921. Systems of mating. V. General considerations. *Genetics* **16**: 97-159

Wright, S. 1929. The evolution of dominance. *Am. Nat.* **63**: 556-561

Wright, S. 1970. Random drift and the shifting balance theory of evolution. In: K. Kojima (ed.), *Mathematical Topics in Population Genetics*. Springer-Verlag, New York

Chapter 5

Barton, NH., D.E.G. Briggs, J.A. Eisen, D.B. Goldstein & N.H. Patel. 2007. *Evolution.* Cold Spring Harbor Laboratory Press.

Clarke, B. & J. Murray. 1969. Ecological genetics and speciation in land snails of the genus *Partula. Biol. Linn. Soc.* **1**: 31-42

Davison, A., N. Constant, H. Tanna, J. Murray & B. Clarke. 2009. Coil and shape in *Partula suturalis*: the rules of form revisited. *Heredity* **103**: 268-278

Grande, C. & N.H. Patel. 2009. Nodal signalling is involved in left-right asymmetry in snails. *Nature* **457**(7232): 1007-1011

Johnson, M.S. 1987. Adaptation and the rules of form: Chirality and shape in *Partula suturalis. Evolution* **41**: 672-675

Johnson, M. S., Clarke, B. & Murray, J. 1977 Genetic variation and reproductive isolation in *Partula. Evolution* **31**: 116-126

Johnson, M.S., Murray, J. & Clarke, B. 1986 Allozymic similarities among species of *Partula* on Moorea. *Heredity* **56**: 319-327

Kuroda, R., B. Endo, M. Abe & M. Shimizu, M. 2009. Chiral blastomere arrangement dictates zygotic left-right asymmetry pathway in snails. *Nature* **462**: 790-794

Lipton, C.S. & J. Murray, J. 1979. Courtship of land snails of the genus *Partula. Malacologia* **19**: 129-146

Millstein, R.L. 2009. Concepts of drift and selection in "the Great Snail Debate" of the 1950s and early 1960s. In: Cain, J. & M. Ruse (eds). *Descended from Darwin: insights into the history of evolutionary studies, 1900-1970*. American Philosophial Society, Philadelphia.

Murray, J. & B. Clarke. 1966. The inheritance of polymorphic shell characters in Partula (Gastropoda). *Genetics* **54**: 1261-1277

Murray, J. O.C. Stine & M.S. Johnson. 1991. The evolution of mitochondrial DNA in *Partula*. *Heredity* **66**: 93-104

Murray, J. & Clarke, B. 1968 Partial reproductive isolation in the genus *Partula* (Gastropoda) on Moorea. *Evolution* **22**: 684-698

Murray, J. & Clarke, B. 1976a Supergenes in polymorphic land snails. I. *Partula taeniata*. *Heredity* **37**: 253-269

Murray, J. & Clarke, B. 1976b Supergenes in polymorphic land snails. II. *Partula suturalis*. *Heredity* **37**: 271-282

Murray, J. & Clarke, B. 1980 The genus Partula on Moorea: speciation in progress. *Proc. R. Soc. Lond. B* **211**: 83-117

Page. L.R. 2003. Gastropod ontogenetic torsion: developmental remnants of an ancient evolutionary change in body plan. *J. Exp. Zool. B. Mol. Dev. Evol.* **297**(1): 11-26

Schwabl, G. & J. Murray. 1970. Electrophoresis of proteins in natural populations of *Partula* (Gastropoda). *Evolution* **24**: 424-430

Scvortzoff, E. 1966. *Chromosome numbers of the land snails of the genus Partula that inhabit the island of Moorea*. M.A. thesis, University of Virginia, Charlottesville

Shimizu, K., M. Iijima, D.H.E. Setiamarga, I. Sarashina, T. Kudoh, T. Asami, E. Gittenberger & K. Endo. 2013. Left-right asymmetric expression of *dpp* in the mantle of gastropods correlates with asymmetric shell coiling. *EvoDevo* **4**: 15

Chapter 6

Allen, J.A. 1884. Zoological Nomenclature. *Auk* (**1884**): 338-353

Anon. 2006. *Starlings and snails at Pasir Panjang hill*. http://www.besgroup.org/2006/01/05/starlings-and-snails-at-pasir-panjang-hill/

Clarke, B. &. J. Murray. 1969. Ecological genetics and speciation in land snails of the genus *Partula*. *Biol. J. Linn. Soc. Lond.* **1**: 31-42

Cuvier G.L. 1798. *Tableau élementaire de l'Histoire Naturelle des Animaux*. Baudouin, Paris

Goodacre, S.L. 2001. Genetic variation in a Pacific Island land snail: population history versus current drift and selection *Proc. R. Soc. B: Biol. Sci.* **268**(1463): 121-126

Goodacre, S.L. 2002. Population structure, history and gene flow in a group of closely related land snails: genetic variation in *Partula* from the Society Islands of the Pacific. *Mol. Ecol.* **11**(1): 55-68

Goodacre, S.L. & C.M. Wade. 2001a. Molecular evolutionary relationships between partulid land snails of the Pacific. *Proc. R. Soc. B: Biol. Sci.* **268**(1462): 1-7

Goodacre, S.L. & C.M. Wade. 2001b. Patterns of genetic variation in Pacific island land snails: the distribution of cytochrome b lineages among Society Island *Partula. Biol. Journ. Linn. Soc.* **73**(1): 131-138

Johnson, M.S., J. Murray & B.C. Clarke. 1986. An electrophoretic analysis of phylogeny and evolutionary rates in the genus *Partula* from the Society Islands. *Proc. R. Soc. Lond. B* **227**(1247): 161-177

Johnson, M.S., J. Murray & B. Clarke. 1986. High genetic similarities and low heterozygosities in land snails of the genus *Samoana* from the Society Islands. *Malacologia* **27**: 97-106

Keulemans, J.G. 1881. *Catalogue of the Bird in the British Museum.* **5**. Trustees of the British Museum, London

Lee, T., J.B. Burch, Y. Jung, T. Coote, P. Pearce-Kelly & D. Ó Foighil. 2007a. Tahitian tree snail mitochondrial clades survived recent mass-extirpation. *Current Biology* **17**: R502-R503.

Lee, T., J.B. Burch, T. Coote, B. Fontaine, O. Gargominy, P. Pearce-Kelly & D. Ó Foighil. 2007b. Prehistoric inter-archipelago trading of Polynesian tree snails leaves a conservation legacy. *Proc. R. Soc. Lond. B* **272**: 2907-2914.

Lee, T., J.-Y. Meyer, J.B. Burch, P. Pearce-Kelly & D. Ó Foighil. 2008. Not completely gone: two partulid tree snail species persist on the highest peak of Raiatea, French Polynesia. *Oryx* **42**: 615-619

Ó Foighil, D 2009. Conservation status update on Society Island Partulidae. *Tentacle* **17**: 30-35.

Paterson, H.E.H. 1985. The recognition concept of species. In: Vrba, E.S. (ed.). *Species and Speciation.* Transvaal Museum Monographs **4**, Pretoria.

Ridgway, R. 1879. On the use of trinomials in zoological nomenclature *Bull. Nutt. Orn. Club.* **4**(3): 129-134

Rothschild, M. 1983. *Dear Lord Rothschild.* Birds, Butterflies and History. Hutchinson, London.

Rothschild, W., & K. Jordan. 1895. A revision of the Papilios of the eastern hemisphere, exclusive of Africa. *Novitat. Zool.* **2**: 167-463

Rothschild, W. 1912. On the term "subspecies" as used in systematic zoology. *Novit. Zool.* **19**: 135-136

Steadman, D.W. 1989. A new species of starling (Sturnidae, *Aplonis*) from an archaeological site on Huahine, Society Islands. *Notornis* **36**: 161-169

Stresmann, E. 1936. The Formenkreis-theory. *Auk* **53**: 150-158

Vagvogli, J. 1975. Body size, aerial dispersal, and origin of the Pacific land snail fauna. *Syst. Zool.* **24**(4): 465-488

Chapter 7

Aberdeen, S. 2013. *The history and future of island conservation in a snail shell. Utilising historically formed baselines and extinction risk analysis within island ecosystems to inform risk assessment and future* Partula *reintroduction planning*. MSc thesis, Imperial College, London

Anon. 1914. *Bulletin de la Société nationale d'Acclimatation* **61**: 82, 403

Gould, S.J. 1993. *Eight little piggies: reflections in natural history*. Norton, New York

Hanna, G.D. 1966. Introduced molluscs of western North America. *Occ. Pap. Calif. Acad. Sci.* **48**: 1-108

Murray, J., E. Murray, M.S. Johnson & B. Clarke. 1988. The extinction of *Partula* on Moorea. *Pacific Science* **42**: 150-153

Pointier, J.P. & C. Blanc. 1995. *Achatina fulica* en Polynésie Francaise. Répartition, caracterisátion des populations et conséquences de l'introduction de l'escargot predateur *Euglandina rosea* en 1982-1983 (Gastropoda - Stylommatophora, Achatinacea). *Malak. Abhand.* **11**: 1-15

U.S. Fish and Wildlife Service. 1992. *Recovery Plan for the O'ahu Tree Snails of the Genus* Achatinella. U.S. Fish and Wildlife Service, Portland, Oregon

Vignal, L. 1915 Quelques observations sur les Glandina guttata. *Bull. Soc natn accl Fr* **62**: 344-349

Chapter 8

Partula Species Management Programme. 2013. *1 March 2013 international meeting report*.

Pearce-Kelly, P., D. Clarke, M. Robertson & C. Andrews. 1991. The display, culture and conservation of invertebrates at London Zoo. *International Zoo Yearbook* **30**: 21-30

Tonge, S. & Q. Bloxham. 1991. A review of the captive-breeding programme for Polynesian tree snails *Partula* spp. *International Zoo Yearbook* **30**: 51-59

Wells, S.M. 1995. The extinction of endemic snails (genus *Partula*) in French Polynesia: is captive breeding the only solution? In: Kay, A. (ed.) *The Conservation Biology of Molluscs*. Occasional Paper of the IUCN Species survival Commission No. **9**. IUCN

Chapter 9

Meyer, J.Y. & J.-P. Malet. 1997. *Study and management of the alien invasive tree* Miconia calvescens *DC. (Melastomataceae) in the Islands of Raiatea and Tahaa (Society Islands, French Polynesia): 1992-1996.* Honolulu (HI): Cooperative National Park Resources Studies Unit, University of Hawaii at

Manoa, Department of Botany. PCSU Technical Report 111.

Chapter 10

Kenis, M., H.E. Roy, R. Zindel & M.E.N. Majerus. 2008. From Biological Control to Invasion: the Ladybird *Harmonia axyridis* as a Model Species. *BioControl* **53**(1): 235–252

Koch, R.L. 2003. The multicolored Asian lady beetle, *Harmonia axyridis*: A review of its biology, uses in biological control, and non-target impacts. *Journal of Insect Science* **3**: 32

Lombaert, E., T. Guillemaud, J.-M. Cornuet, T. Malausa, B. Facon & A. Estoup. 2010. Bridgehead Effect in the Worldwide Invasion of the Biocontrol Harlequin Ladybird. *PLoS ONE* **5**(3): e9743

Meyer, J.-Y. 1996. Status of *Miconia calvescens* (Melastomataceae), a dominant invasive tree in the Society Islands (French Polynesia). *Pacific. Sci.* **50**: 66-76

Meyer, J.-Y. 2009. The *Miconia* Saga: 20 Years of Study and Control in French Polynesia (1988-2008). In: Lope, L.L., J.-Y. Meyer & B.D. Hardesty (eds.). *Proc. Inernl. Miconia Conference, Hawaii 2009.*

Meyer, J.-Y. & M. Fourdrigniez. 2011. Conservation benefits of biological control: The recovery of a threatened plant subsequent to the introduction of a pathogen to contain an invasive tree species. *Biol. Conserv.* **144**: 106–113

Meyer, J.-Y., M. Fourdrigniez & R. Taputuarai. 2012. Restoring habitat for native and endemic plants through the introduction of a fungal pathogen to control the alien invasive tree *Miconia calvescens* in the island of Tahiti. *BioControl* **57**: 191-198

Chapter 11

Coote, T. 2002. Protecting the last populations of endemic tree snails in the Society Islands, French Polynesia. *Tentacle* **10**: 20

Cunningham A.A. & P. Daszak. 1998. Extinction of a species of land snail due to infection with a microsporidian parasite. *Conserv. Biol.* **12**(5): 1139-1141

Pearce-Kelly, P., & D. Clarke (eds) 1996. *Partula 1995: London: Zoological Society London*

Pearce-Kelly, P., G.M. Mace & D. Clarke. 1995.The release of captive bred snails (*Partula taeniata*) into semi natural environment. *Biodiv. Conserv.* **4**: 645-663

Chapter 12

Coote, T. 2002. Protecting the last populations of endemic tree snails in the Society

Islands, French Polynesia. *Tentacle* **10**: 20
Coote, T. 2007. Partulids on Tahiti: differential persistence of a minority of endemic taxa among relict populations. *Am. Malacol. Bull.* **22**: 83-87
Coote, T. 2012. Tahiti become the focus of *Partula* conservation. *Tentacle* **20**: 28030
Coote, T. & W. Teamotuaitau. 2005. Partulids on Tahiti: an interesting distribution among surviving populations. http://www.malacological.org/meetings/archives/2005/documents/Abstract_volume-03.html
Coote, T., D. Clarke, C.S. Hickman, J. Murray & P. Pearce-Kelly. 2004. Experimental release of endemic *Partula* species, extinct in the wild, into a protected area of natural habitat on Moorea. *Pacific Science* **58**: 429-434
Coote, T., E. Loève, J.-Y. Meyer & D. Clarke. 1999. Extant populations of endemic partulids on Tahiti, French Polynesia. *Oryx* **33**: 215-222
Lee, T., J.B. Burch, Y. Jung., T. Coote, P. Pearce-Kelly & D. Ó Foighil. 2007. Tahitian tree snail mitochondrial clades survived recent mass-extirpation. *Current Biology* **17**: R502-R503
Lee, T., J.-Y. Meyer, J.B. Burch, P. Pearce-Kelly & D. Ó Foighil. 2008. Not completely gone: two partulid tree snail species persist on the highest peak of Raiatea, French Polynesia. *Oryx* **42**: 615-619
Lee, T., J.B. Burch, T. Coote, P. Pearce-Kelly, C. Hickman, J-Y. Meyer & D. Ó Foighil. 2009. Moorean tree snail survival revisited: a multi-island genealogical perspective. *BMC Evol. Biol.* **9**(204): 1-16.
Ó Foighil, D. 2009. Conservation status update on Society Island Partulidae. *Tentacle* **17**: 30-35

Chapter 13

Anonymous. 2000. Flatworm (*Platydemus manokwari*) in Samoa. *SAPA Newsletter* **2**/00: 3-4
Arrhenius, S. 1896. On the influence of carbonic acid in the air upon the temperature of the ground. *Philosophical Magazine* **41**: 237-76
Arrhenius, S. 1908. *Worlds in the Making*. New York: Harper & Brothers.
Avagliano E. & J.N. Petit. 2009. *Etat des lieux sur les enjeux du changement climatique en Polynésie française*. Ministère de l'Environnement de la Polynésie française, Direction de l'Environnement de la Polynésie française, Station Gump, UC Berkeley.
Eldredge, L.G. & B.D. Smith. 1994. Introductions and transfers of the triclad flatworm *Platydemus manokwari*. *Tentacle* **4**: 8
Eldredge, L.G. & B.D. Smith. 1995. Triclad flatworm tours the Pacific. *Aliens* **2**: 11
Hopper, D.R. & B.D. Smith. 1992. The status of tree snails (Gastropoda: Partulidae) on Guam, with a resurvey of sites studied by H.E. Crampton in 1920.

Pacific Science **46**:77-85

Intergovernmental Panel on Climate Change. 2013. *Climate Change 2013: the physical science basis*. http://www.climatechange2013.org/images/uploads/WGIAR5_WGI-12Doc2b_FinalDraft_All.pdf

Justine, J., L. Winsor, D. Gey, P. Gros, J. Thévenot. 2014. The invasive New Guinea flatworm *Platydemus manokwari* in France, the first record for Europe: time for action is now. *PeerJ* **2**: e297 http://dx.doi.org/10.7717/peerj.297

Kerr, A.M. 2013. *The partulid tree snails (Partulidae: Stylommatophora) of the Mariana Islands, Micronesia*. University of Guam Marine Laboratory Technical Report **153**

Meyer, J.-Y. 2010. Montane cloud forests on remote islands of Oceania: the example of French Polynesia (South Pacific Ocean). In: Bruijnzeel, L.A., F.N. Scatena & L.S. Hamilton (eds.) *Tropical Montane Cloud Forests: Science for Conservation and Management*. Cambridge University Press.

Muniappan, R. 1987. Biological control of the giant African snail, *Achatina fulica* Bowdich, in the Maldives. *FAO Plant Protection Bulletin* **35**(4): 127-133

Muniappan, R. 1990. Use of the planarian, *Platydemus manokwari*, and other natural enemies to control the giant African snail. In: *The use of natural enemies to control agricultural pests. Proceedings of the international Seminar "The use of parasitoids and predators to control agricultural pests" Tsukuba, Japan, October 2-7, 1989*. Food and Fertilizer Technology Canter for the Asian and Pacific Region, Taipei. pp.179-183

Muniappan, R., G. Duhamel, R.M. Santiago & D.R. Acay. 1986. Giant African snail control in Bugsuk island, Philippines, by *Platydemus manokwari*. *Oléagineux* **41**(4): 183-186

Peterson, T. C., M.P. Hoerling, P.A. Stott & S. Herring (eds.) 2013. Explaining Extreme Events of 2012 from a Climate Perspective. *Bull. Amer. Meteor. Soc.* **94** (9): S1-S74

Pouteau, R., J.-Y. Meyer, R. Taputuarai & B. Stoll. 2010. La fonte de la biodiversité dans les îles: modélisation de l'impact du réchauffement global sur la végétation orophile de Tahiti (Polynésie française). *Vertigo* **10**(3) http://vertigo.revues.org/10580

Slocum, G. 1955. Has the amount of carbon dioxide in the atmosphere changed significantly since the beginning of the twentieth century? *Monthly Weather Review* (**October 1955**): 225-232

Smith, B.D. 1995. *Status of the Endemic Tree Snail Fauna (Gastropoda: Partulidae) of the Mariana Islands*. Prepared for U.S. Fish and Wildlife Service, Pacific Islands Ecoregion, Honolulu, HI.

Smith, B.D. 2013. *Taxonomic inventories and assessments of terrestrial snails on the islands of Tinian and Aguiguan in the Commonwealth of the Northern*

Mariana islands. University of Guam Marine Laboratory Technical Report **154**

Smith, B.D., & D.R. Hopper. 1994. The Partulidae of the Mariana Islands: Continued threats and declines. *Hawaiian Shell News* **42**(6):10-11

Glossary

alleles – different versions of a particular gene
allopatric species – species that live in different areas
apostatic selection – selection by a predator whereby the rarest form of its prey is favoured
allozyme (or alloenzyme) - different forms of the same enzyme
binomial system – the system of giving two names to a species: its genus and species names
biological control – control of pest species using natural enemies (predators, parasites or diseases)
biological species concept – the definition of a species as a group of individuals capable of interbreeding, separated from other such groups
chirality – the direction of shell coiling, either clockwise (right-handed or dextral) or anti-clockwise (left-handed or sinistral)
clade – a branch on an evolutionary tree
cladisitcs – the branching pattern of species origin seen in an evolutionary tree
conchologists – snail biologists and shell collectors
cryptic species – hidden species, not visibly distinct and identified as separate species based only on their DNA
dextral - right-handed, in snail a shell that spirals in a clockwise direction
dominant allele – a form of a gene that is always expressed, irrespective of with which other alleles it is paired
electrophoresis – the method of separating enzymes or DNA fragments based on their size by passing an electric current though a gel on which the chemical sample has been placed
Evolutionarily Significant Units (ESU) – groups of individuals that are of significance in evolution, may be populations, or smaller groups, not necessarily species
filial generation – the offspring of a controlled mating,, such as a Mendelian crossing
genetic drift – random genetic change, caused by chance rather than by selection
genus – a grouping of closely related species
haplotypes – different versions of mitochondrial genes
heterozygote – an individual with both dominant and recessive alleles for a particular gene
homozygous dominant – an individual with only dominant alleles for a particular gene
homozygous recessive – an individual with only recessive alleles for a particular gene

'maternal effect' - inheritance of characteristics determined by the proteins in the egg., rather than the genes of the individual

Mendelian crosses – matings between individuals designed to determine which genetic alleles are dominant or recessive

Mendelian genetics – inheritance of characteristics determined by the interaction dominant and recessive genetic alleles

microsporidians – microscopic single-celled parasitic fungi

mitochondria – the small structures within a cell responsible for much of the cell's metabolism, contains its own strand of DNA

mitochondrial DNA – the single strand of DNA contained within each mitochondrion

'modern synthesis' – the synthesis of Darwin's theory of evolution by natural selection with the science of genetics

nomenclature – the system of naming species, giving them a 'Latin' name, a component of taxonomy

nucleus – the part of the cell containing the DNA inherited from both parents

phylogenetic distance – a measure of how different two species are in their evolutionary history (e.g. time since separation, amount of genetic difference)

phylogeny – an evolutionary tree

polymorphism – different forms of the same species, such as banding patterns

radula – a snail's feeding structure composed of a tongue-like structure covered in rows of hundreds of microscopic teeth

recessive allele – a form of a gene that is only expressed when it is not paired with a more dominant form

restriction fragment length polymorphisms (RFLP) – separation of different versions of the same DNA sequence through electrophoresis

sinistral - left-handed, in a snail a shell that spirals in an anti-clockwise direction

species concepts – different ways of defining a species, usually based on the 'biological species concept'

subspecies – a subdivision of a species, a local variety

supergene – several genes that are inherited together, such that they can be considered a single entity

sympatric species – species that live in the same place

taxonomy – the classification of living organisms by a hierarchy of species, genera, families, orders etc.

torsion – the 180° rotation that occurs in a snail's development, twisting the shell round so that it lies over the snail's head

Index to species

A

Achatina 97-99, 101, 103-104, 113, 116-117, 119-120, 126-127, 138, 148, 150
Achatinella 39, 49-50, 101, 103
affinis 27, 89, 91, 98, 103, 108, 138, 143-144, 146, 152
arguta 37, 43, 103, 108, 137, 139-140
assimilis 88
attentuata 62
aurantia 65, 74, 78, 80-81, 83, 105, 107, 134, 139
australis 25, 29-30

B

bellula 151
bicolor' 38
Bradybaena 119-120
burchi 68, 90, 143-144

C

canalis 74
Cepaea 55-56, 69-71
clara 84, 91-92, 103, 105, 107, 143-144, 152, 154
conica 74
cytherea 74

D

dentifera 62, 92, 111, 121-122, 139, 142
dryas 151

E

Edentulina 98, 100
elongata 65
Eua 66, 88-89
Euglandina 98-104, 108-113, 115-117, 119-122, 125-135, 137-140, 143-146, 148-152, 154, 161
exigua 62, 78, 81-82, 97, 101-102, 104-105, 134, 139

F

faba 1, 21-23, 25-26, 27-31, 33, 43, 92, 95, 108-109, 119-122, 124, 126, 134, 139-140, 142, 145, 154
filosa 62, 91
formosa 62
fragilis 30

G

ganymedes 151
garretti 120
gibba 30, 37-38, 74, 88, 108, 139, 149-150
Gonaxis 98-100, 149-150

H

hamadryas 151
hebe 89, 109, 119, 134, 139
hyalina 91-92, 103, 105, 107-108, 138-139, 142-144, 146, 154

I

incrassa 84, 91-92

J

jackieburchi 68, 135, 157

L

labrusca 124, 139-140, 145
Lissachatina 97
lutea 20, 30-31, 44, 88, 108
Lymnaea 61, 76

M

meyeri 92, 146-148, 152
mirabilis 60, 65, 76, 81, 93, 105, 108, 139
mooreana 62, 65, 74, 78, 81, 83, 93, 125, 139-140

N

nodosa 62, 74, 82, 89, 91, 107-108, 139, 142, 146

O

olympia 74, 78, 83
oreas 151
otaheitana 25-27, 30-31, 47, 56, 58, 61-62, 68, 74, 82-83, 91, 94, 103, 112-113, 139, 143-144, 152

P

Platydemus 148-150
producta 91, 125, 130
pudica 30

R

radiata 37
radiolata 37-38, 88, 139-140
rosea 103, 108-109, 112, 139-140

S

Samoana 30, 62, 66, 68, 74, 89-90, 105, 124, 135, 138-140, 143-146, 151-152
stevensonia 74
strigata 151
suturalis 60, 62, 65, 69-70, 73-75, 77-84, 89, 94, 105, 108, 137, 139

T

taeniata 31, 62, 65, 70-74, 78, 80-82, 92-94, 105, 108, 135, 137, 139-140, 145, 152
tohiveana 62, 74, 76, 78, 80-81, 83-84, 93, 105, 108, 137, 139
tristis 89, 92, 109, 139, 142
turgida 65, 88-89, 92, 109, 134, 139-140

V

varia 94, 103, 108-109, 137, 139-140
vexillum 65

Index to people

A

Agassiz 40, 42, 47
Anderson 19-20
Anton 30
Arrhenius 151-152

B

Bailey 71, 80
Banks 2-3, 5-11, 15, 19-20, 25, 95, 124
Barnard 40, 51
Bell 151
Bentinck 23
Bloxham 105, 161
Bougainville 2, 5, 12, 14
Brazier 37-38, 157
Broderip 33, 35, 39
Bruguière 29-30, 74
Burch 68, 90-92, 101

C

Cain 69, 70
Callendar 151
Cassin 84
Chamberlin 98
Chemnitz 25-29, 33
Christian VIII of Denmark 27
Clarke, Bryan 69-72, 76, 80, 83-84, 87-88, 91, 93-94, 96, 101-103, 112, 120, 136
Clarke, Dave 107-8, 137, 139-140, 143
Clench 42, 66-67
Commerçon 2
Cook 1, 6-9, 11-12, 17, 19, 20-23, 25-27, 29-30, 92, 94, 111, 124, 132,
Cooke 26, 56, 65-66, 68, 109
Coote 137-140, 143-144, 146, 154
Correns 54-55
Cracknell 137
Crampton 49-51, 53, 55-62, 64-68, 70-72, 74-76, 80, 82-84, 87, 91, 109, 120, 144, 153
Cuming 31-35, 39, 41, 44, 46
Cuvier 86

D

Darwin 2, 41, 49-50, 52, 54, 74, 84-86, 157-158, 167
De Vries 54, 55

E

Ellis 3, 5

F

Fabricius 3
Férussac 30
Forster 11, 15, 17, 19, 20, 94-95, 132
Freycinet 30

G

Gaimard 30
Garrett 35, 38-44, 46-47, 49, 56, 64-65, 74, 84, 109, 119-120
Gaudichaud-Beaupré 30
Gmelin 29, 43, 95
Goodacre 88-89, 93-94, 139
Gray 31, 33-35
Gulick 49-51, 66, 157

H

Haalelea 39, 41
Hartert 85
Hartman 38

Hawkesworth 6
Hickman 145
Humphrey 23, 25, 30

J

Johnson 69, 71-72, 77, 80, 83, 84, 88, 91, 96, 107, 136
Jordan 50, 85

K

Kamehameha I 39
Klocek 108, 137
Kondo 66-68, 88, 90, 98, 99, 101, 150

L

Lambert 144
Lang 55-56
Lee 90
Lesson 30, 37, 44
Ligny 98, 148
Linnaeus 3, 4, 5, 11, 21, 27, 29, 86
Lipton 78

M

Mace 134
Malet 119, 123, 161
Martyn 20-23, 25, 27, 29-30, 95
Maskelyne 1
Mayer 47-49, 56, 61, 65
Mayr 85-86
Mead 98-99
Melvill 34
Mendel 54-55, 58, 72
Meyer 123, 130, 139, 145-146
Montague 6, 56
Muniappan 148-149
Murray 69-72, 76, 78, 80, 83-84, 87-88, 91, 96, 101-103, 106-108, 136-137, 145

N

Nägeli 54

O

Omai 19
Opoony 10
Otari 42
Owen 34, 98
Ó Foighil 90, 92

P

Pearce-Kelly 60, 107, 112, 137-138, 143
Pease 38-44, 46, 49, 83-84, 119, 157
Peterson 98-99, 164
Pfeiffer 33, 35, 39
Pilsbry 66
Portland, Duchess of 20, 23-24

Q

Quoy 30

R

Ray 4, 47
Rossiter 37-38
Rothschild 85

S

Schlegel 84
Sheppard 69-70
Smith 150
Sparman 11-12
Spencer 137

Spengler 25-27
Sturtevant 61-62

T

Tonge 105
Tucker Abbott 98
Tupia 9-10
Tyron 66

W

Wade 88-89
Wagner 50
Wallace 21, 49, 52
Wallis 2, 12
Wright 50-51

www.ingramcontent.com/pod-product-compliance
Lightning Source LLC
Chambersburg PA
CBHW052023290426
44112CB00014B/2350